ENERGY AND ENVIRONMENT REGULATION

STUDIES IN REGULATION

General Editor: George Yarrow, Director, Regulatory Policy Institute and Regulatory Policy Research Centre, Hertford College, Oxford

Government regulation of business activity is a pervasive characteristic of modern economies, including those most committed to free markets. For good or ill regulation has far reaching implications for economic performance, and understanding the processes at work is an important task for anyone seeking to analyse the determinants of performance. Quite frequently, however, analysis is restricted to a very specific aspect of business activity or to a particular sector of the economy, an approach that serves to limit the insights into regulatory issues that may be gained.

A guiding principle behind this series of books is that regulatory processes exhibit a number of common features that are likely to manifest themselves in a range of different circumstances. A full understanding of the motives for and effects of regulation therefore requires study of these common features, as well as the specifics of particular cases of government interventions. Thus, it is possible to learn something relevant about, say, the regulation of utilities from the study of financial services regulation, or about industrial policy from the study of environmental regulation.

This focus on regulatory processes in general, as well as on specific aspects of particular interventions, also points to the value of interdisciplinary analysis. Policy formulation, development and implementation each have political, legal and economic aspects, and the boundaries between traditional academic disciplines can be obstacles to progress in regulatory studies. In this series, therefore, a wide range of different perspectives on regulation and on regulatory processes will be presented, with the aim of contributing to the development of new insights into important policy issues of the day.

Energy and Environment Regulation

Edited by

Helen Lawton Smith
Director of Science Policy Studies
Regulatory Policy Research Centre
Hertford College
Oxford

and

Nick Woodward
Director of Environmental Policy Studies
Regulatory Policy Research Centre
Hertford College
Oxford

First published in Great Britain 1996 by
MACMILLAN PRESS LTD
Houndmills, Basingstoke, Hampshire RG21 6XS
and London
Companies and representatives
throughout the world

A catalogue record for this book is available
from the British Library.

ISBN 978-1-349-25059-2 ISBN 978-1-349-25057-8 (eBook)
DOI 10.1007/978-1-349-25057-8

First published in the United States of America 1996 by
ST. MARTIN'S PRESS, INC.,
Scholarly and Reference Division,
175 Fifth Avenue,
New York, N.Y. 10010

ISBN 978-0-312-15951-1

Library of Congress Cataloging-in-Publication Data
Lawton Smith, Helen.
Energy and environment regulation / Helen Lawton Smith and Nick Woodward.
p. cm.
Includes bibliographical references and index.
ISBN 978-0-312-15951-1
1. Energy industries—Environmental aspects. 2. Energy industries—Environmental aspects. 3. Environmental law.
I. Woodward, Nick.
HD9502.A2L396 1996
363.7'07—dc20 96–11320
CIP

10 9 8 7 6 5 4 3 2 1
05 04 03 02 01 00 99 98 97 96

CONTENTS

CONTRIBUTORS

Brenda Boardman Senior Research Fellow, Environmental Change Unit, University of Oxford; PowerGen Fellow in Energy Efficiency, St Hilda's College, Oxford

Roger Burritt Senior Lecturer in Accounting, Department of Commerce, the Australian National University, Canberra, Australia

Peter Carter Deputy Director General OFFER

Nick Eyre Eyre Energy Environment

Kazuya Fujime Director for Research, The Institute of Energy Economics, Japan

Leigh Hancher Professor of Public Economic Law, Faculty of Law, Erasmus University of Rotterdam, the Netherlands

Nicholas Hartley Head of Energy, Environmental Economics and Statistics Division, Department of Trade and Industry, UK

Akihiro Kuroki Senior Researcher, National Institute for Research Advancement, Japan

Helen Lawton Smith Research Director, Regulatory Policy Research Centre, Hertford College, Oxford

David Maddison Research Fellow, Centre for Social and Economic Research on the Global Environment (CSERGE), University College, London

Mike Parker Ex-Chief Economist, British Coal; Honorary Fellow, Science Policy Research Unit, University of Sussex

Wolfgang Pfaffenberger Professor of Economics (Economic Policy), Universitat Oldenburg Institut for Volkswirtschaftslehre, Oldenburg, Germany

Larry E Ruff Managing Director, Putnam, Hayes & Bartlett

Philippe Sands Barrister; Lecturer in Law, School of Oriental and African Studies, London University; Legal Director, Foundation for International Environmental Law and Development

Jim Skea British Gas/ESRC Fellow, Science Policy Research Unit, University of Sussex

Nick Woodward Fellow, Templeton College, Oxford

Kenji Yamaji Professor, Department of Electrical Engineering, Universit of Tokyo, Japan

George Yarrow Director, Regulatory Policy Research Centre and Fellow, Hertford College, Oxford

ABBREVIATIONS

ALARA	As Low As Reasonably Achievable
AmUJILP	American University Journal of International Law and Policy
AOSIS	Alliance of Small Island States
BATNEEC	Best Available Technology Not Entailing Excessive Cost
BAU	Business As Usual
BCC	British Coal Corporation
BPEO	Best Practicable Environmental Option
BTUs	British Thermal Units
CAA	Clean Air Act
CCGT	Combined Cycle Gas Turbine
CEC	Commission of the European Community
CETA	Carbon Emissions Trajectory Assessment
CFCs	Chlorofluorocarbons
CHP	Combined Heat and Power
ColJIELP	Colorado Journal of International Environmental Law
CRIEPI	Central Research Institute of Electric Power Industry
CVM	Coningent Valuation Method
DGES	Director General of Electricity Supply
DSM	Demand Side Management
EC	European Commission
ECS	Economic and Social Committee
EIS	Energy and Industry sub-group
EPA	Environmental Protection Agency
EQS	Environmental Quality Standards
FGD	Flue Gas Desulphurisation
GEF	Global Environmental Facility
GHGs	Greenhouse Gases
GNP	Gross National Product
HMIP	Her Majesty's Inspectorate of Pollution
IBRD	International Bank for Reconstruction and Development
ICAO	International Civil Aviation Organisation
IER	International Environment Reporter
IGCC	Integrated Gasification Combined Cycle
ILM	International Legal Materials
INC/FCCC	Intergovenmental Negotiating Committee for a Framework Convention on Climate Change

IPC	Integrated Pollution Control
IPCC	Intergovernmental Panel on Climate Change
kWh	Kilowatt-hours
LCP	Large Combustion Plant
LNG	Liquid Nitrogen Gas
LRTAP	Long-range Transboundary Air Pollution
MITI	Ministry of International Trade and Industry
MMBTUs	Million British Thermal Units
NGC	National Grid Company
NOx	Nitrogen Oxides
NSPS	New Source Performance Standard
OECD	Organisation for Economic Co-operation Development
OFFER	Office of Electricity Supply
OFGAS	Office of Gas Supply
OJ	Official Journal of the European Committee
Oz	Tropospheric Ozone
PCBs	Polychlorinated BiPhenyls
PFBC	Pressurised Fluidised Bed Combustion
PUCs	Public Utility Commissions
RD&D	Research, Development and Demonstration
RECS	Regional Electricity Companies
RECIEL	Review European Community and International Envionmental Law
SCR	Selective Catalytic Reduction
SIPs	State Implementation Plans
SOX	Sulphur Dioxides
TEU	Treaty on European Union
TPA	Third Party Access
TSO	Transmission System Operator
UNCED	United Nations Conference on Environment and Development
UNECE	United Nations Economic Commission for Europe
UNDP	United Nations Development Programme
UNEP	United Nations Environmental Programme
UNTS	United Nations Treaty Series
USOTA	United States Office of Technology Assessment
WMO	World Meterological Organisation

PREFACE

The chapters in this book are mainly revised versions of papers given at an international workshop on "Energy and Environment Regulation" held at Hertford College Oxford September 1993. The purpose of the workshop was to "further the debate on the relationship between the regulation of energy production and environmental regulation in the global context".

The workshop brought together people with widely differing perspectives. Participants came from many countries including Japan, the USA, Germany, Poland, Australia, France as well as the UK. They included academics from a number of disciplines (economics, science policy, law, geography, management), research institutions specialising in the study of energy, industrialists, policy makers, and consultants. The synergy generated by this diversity contributed significantly to the academic success of the workshop. This book reflects the achievements of the workshop.

The first chapter of the book, by Nick Woodward, is intended to provide an environmental context for energy regulation. The chapter starts by sketching the background to environmental concern by discussing the evidence of trends and some characteristic problems, particularly those which make them intractable to conventional modes of analysis and solution. He then reviews energy in the light of these trends, and follows this by considering some alternative perceptions which underpin approaches to the formulation, analysis and solution of environmental issues, and which underlie many of the debates and differences. He concludes that the evidence indicates the complex interdependence of environmental trends, a consensus on their direction and a need for political resolution.

The rest of the book follows the same format as the workshop. The papers are divided into four sections, using the four themes from the workshop. They are followed by the comments from the discussants of each session at the workshop.

Theme I Assessing Regulatory Impacts

Nick Eyre's paper discusses the setting of objectives for environmental regulation. Although a simple definition of environmental regulation is that of regulation undertaken to protect environmental standards, the precise objectives of environmental regulation are frequently less clear. He makes the point that in the absence of clear overall policy guide-lines, regulatory practice is often largely the outcome a process of negotiation between polluter and regulator.

Eyre suggests that environmental regulation is undertaken with a range of different objectives. He identifies four broad approaches: acceptable costs; adequate safety and standards; economic efficiency; and sustainable development. He then discusses the problems associated with measuring benefits. He concludes that sustainable development provides a useful goal but has yet to be converted into concrete policy and regulatory objectives.

David Maddison's chapter illustrates the problems of estimating the true dimensions of climate change. He takes one method, cost-benefit analysis, and explores what conventional assumptions imply about the dimensions of the climate change problem and reviews the extent to which society ought to incur costs now in order to prevent future climate change. His model incorporates information relating to economic growth assumptions, carbon emission forecasts, abatement cost estimates and global warming damage functions.

A Japanese perspective on the analysis of the effects of environmental regulation is given by Akihiro Kuroki. His chapter takes a case study of counter-measures against sulphur dioxide in Japan. Kuroki argues that as Japanese policy measures against sulphur oxides, carbon monoxide and photochemical smog have been largely successful, these can be used as a model for dealing with pollution problems in other countries. However, major investment costs are required to achieve these objectives, and, moreover, they have a significant impact on the economy as a whole. The case study illustrates the potential benefits and costs of particular policy measures.

Roger Burritt's chapter discusses the approach of a new area of study, environmental accounting, to that of energy regulation. It was first used in national income accounting when it was recognised that GNP had defects as a measure of economic progress if environmental factors were not taken into account. Environmental accounting was developed as an additional form of reporting. Since the late 1980s, environmental accounting at the level of nations has been overtaken by reporting at the corporate level.

Burritt argues that environmental accounting has the potential to become an important tool in the strategy mix available for implementing environmental law, and for evaluating firms' environmental impacts.

Brenda Boardman in her discussant's comments questions whether policy should primarily support the objective of achieving precision in valuing and assessing the externalities of energy production, or should take a broad approach so as to stimulate action as quickly as possible. Her conclusion is that the latter is preferable. Evidence suggests that simple universal targets are influential in changing policies, and that public concern, as illustrated by the Japanese example, provide sufficient basis for action.

Theme II Energy Pricing

The three chapters in this section are written by economists, but from different standpoints. George Yarrow is an academic economist, Nick Hartley is the Head of Energy and Environment Economics at the UK's Department of Trade and Industry. Kazuya Fujime is Director for Research at the Institute of Energy Economics, Japan. The discussant, Mike Parker, is also an economist. He was the chief-economist with British Coal and is now at the University of Sussex.

Nick Hartley's chapter draws together two different developments, which have coincidentally come together. These are firstly, shifts in the structure of the energy industries in the UK, driven by the process of privatisation; and secondly, the changing perception of the value of environmental resources, in particular the recognition of the damage caused by the extraction, processing and use of energy. The central theme is the link between energy pricing and the environmental impacts of energy production.

George Yarrow's chapter examines regulatory policies and energy prices. He discusses the complexity of the UK regulatory regime, and the diverse and conflicting policy agenda. He argues that while economic analyses of public policy tend to stress two aspects: effects on economic efficiency, and effects on the distribution of income or resources, it is very frequently the distributional effects which determine which policy is to be adopted. The UK case is illustrated by the example of the extension of the VAT base. The discussion is then extended to the global warming problem. He concludes that the effective conduct of public policy requires much more than greater scientific understanding of the environmental effects of economic activity and standard methods of economic analysis.

What is needed are better political skills to diagnose the effects of particular policies, and to build appropriate consensuses.

Kazuya Fujime reports on a study which estimated generating costs of different generation systems in Japan. His results indicate the advantages of nuclear energy.

Perhpas not untypically, the economists are not in total agreement. Mike Parker is strongly critical of conventional economic methods of determining energy prices. He refers to the practical difficulties of political decision making discussed by George Yarrow and Nick Hartley, but does not entirely agree with their conclusions. He criticises Nick Hartley's conclusion that environmental problems are economic problems which can be addressed by classic economic remedies, but is more optimistic of the effectiveness of the price mechanism in meeting a particular well specified economic target.

Theme III Public Policy Towards New Investment

This section contains papers from three countries, Japan (Kenji Yamaji), UK (Jim Skea) and the USA (Larry Ruff), and each provide insights into current assumptions and pre-occupations in different countries.

Jim Skea's chapter examines the relationship between investment, technological change and environmental regulation. It focuses particularly on the problem of emissions from energy production and transformation. It uses two basic building blocks. The first is a conceptual model of investment and stock replacement which provides a framework for examining questions about new plant investment, retrofitting and retirement. The second is an examination of different types of regulatory instrument. He draws attention to the interaction between the choice of regulatory instruments, the style of regulation adopted in any one country, the involvement of industry in the creation of regulations, and technology push.

Larry Ruff's argument is that energy conservation by utilities and other industries, will inevitably play a major role in the costly effort to reduce environmental problems. Utilities can facilitate conservation by finding innovative ways to offer conservation services in the market along with energy itself, competing in the market to meet consumer needs (demand) but at minimum economic and environmental cost. The chapter examines the view that utility demand-side programmes offer an environmental free lunch on a large scale. He argues that energy conservation neither requires nor justifies setting aside standard economic principles or processes, and that there are no free lunches.

Nick Woodward examines, in his discussant's comments, what these diverse papers have in common. He concludes that there are three strands of analysis. First, is the recognition that economic agents behave economically, reacting and adjusting to price/cost signals and to regulations over time. Second, they highlight the critical importance of consultation and information exchange. The third is the demonstration of the contextual nature of different countries' conditions -- hence the value of comparative analysis. He then develops the theme of investment in new technologies (or new fuel sources), pointing to the possibilities of further kinds of pollution being generated as a result of specific public policy and to the phenomenon of 'lock-in' (dependence and familiarity with current sources, processes and technologies) which may constrain the options of exploring radically new kinds of technology.

Theme IV Multi-Regulation

This section contains papers by three academics, (two of them specialising in law), and by one practitioner. The complexity of the regulatory framework on energy and environment regulation operating in different countries within the European and wider international contexts is clearly demonstrated by these contributions.

Wolfgang Pfaffenberger's chapter takes the case of multi-regulation in Germany, a country with a reputation for its high standards of environmental regulation. He gives an outline of the various instruments employed under the federal system, and the results achieved. He concludes that the administrative approach to environmental policy which has been developed during the last twenty years has reached its limits. He puts forward the view that the alternative concept to present day regulation is a market approach, this could be achieved by putting a scarcity value on environmental factors by taxation or by restricting their quantities.

The evidence in the chapter supports his statement that while some of the results are impressive, for example in the reduction of sulphur dioxide, the efficiency record is not. The German approach has been unable to deal with the most problematic sector, transport, and cannot cope with carbon dioxide and other greenhouse gases. He also highlights the special problems of unification.

Leigh Hancher's chapter is concerned with the impact of EC rules and regulations on the way in which energy sectors are organised. This issue, achieved by supply-side push in the form of subsidies for R&D to government institutions/industry/universities, or demand-pull, via taxes on existing sources and technologies, deliberately to encourage substitution. Either (or both) options require a more proactive, long-term stance.

We are very grateful to the contributors to this volume. We recognise that we have been extremely fortunate in the co-operation we have received from all of the many authors. We are also indebted to the other participants at the workshop for their stimulating contributions to the debate and additional information.

We would also like to acknowledge the invaluable assistance and support provided to us in the editing of this book. In particular we would like to thank Mike Parker for diligently reading and commenting on the chapters, Nick Woodward for his unfailing support and unstinting editorial efforts, and Susan Belton-Jones for the massive task of preparing the manuscript for the printers.

The financial and logistical support of the following is gratefully acknowledged: Hertford College, and Coopers and Lybrand for supporting the conference.

We would also like to thank Giovanna Davitti of the Macmillan Press for her encouragement and support.

Helen Lawton Smith and George Yarrow,
Regulatory Policy Research Centre, Hertford College.

PART 1

Introduction

CHAPTER 1

The context for energy and environmental regulation

Nick Woodward

Introduction

The frontispiece of a book entitled 'Modern Business: Principles and Practice' (S. Chapman, 1935) shows two chimneys smoking into a sunlit sky, captioned 'Active Symbols of the Modern Factory'. The same book devotes three pages, out of 500, to 'Woman's Place in the Business World', noting (p.496) that 'above all there are a large number of secretarial posts.'. Sixty years on both items seem quaint. Within five years of the book's publication women were manning factories for the war effort: and few people now, whatever their views on tradable permits, would choose smoke-stacks as active and beneficent symbols of energetic manufacturing.

This small example illustrates how views, assumptions and values (perhaps practice) can change through time. Over the past twenty years concern for the environment has impinged on public and political consciousness. Will this concern seem equally quaint in fifty years time?

This chapter sketches the background to this environmental concern - the evidence of trends (Section 1) and some characteristic problems, particularly those which make them intractable to conventional modes of analysis and solution (Section 2). It then reviews energy in the light of these trends (Section 3), and considers (Section 4) some alternative perceptions which underpin (philosophically, psychologically, politically, commercially) approaches to the formulation, analysis and solution of environmental issues, and which underlie many of the debates and differences. These approaches in turn influence the design and functioning of the human systems which are preconditions for confronting, and perhaps resolving, environmental problems.

Environmental trends

There is abundant evidence for the increase in international environmental concern over the last two decades: growth of movements dedicated to action, analysis and lobbying, on general and particular issues; of movements impinging on or adopting political representation; of consumer concern, over health, environmentally benign products, organic food and gardening; and of governmental meetings, legislation, pledges - the Brundtland report, the Rio summit, EEC (now EU) directives, UN initiatives.

There will, however, be more dispute about the nature, extent, and significance of these concerns. At one extreme they may be viewed as a form of moral panic, about matters which markets, science, technology, rationality and nature will sort out in time, in the tradition of self adaptive systems, for such is progress and doomsday has been proclaimed before. And many of the warnings have an emotional tenor which invites rejection by those with a preference for order, structure and stability. As Radford (1990, p.219) puts it - 'since 1970 there has been a series of near-apocalyptic warnings, not just from the 'green' movements, but from national political leaders and from senior scientists and economists' and, 'since 1988, these warnings have become more urgent and more apocalyptic'.

At the other extreme of concern, even now remedial action may be too late - too much has been destroyed and polluted, and many trends are irreversible, except in terms of hundreds of years or of earth's long term capacity to adapt, with or without humanity (cf Lovelock, 1990). This section will outline six major categories of trend, following Holdgate, 1991, p.14 ff.

Population growth

The world's population has increased from 1 bn (1800), 2 bn (1900) to over 5.2 bn in the 1990s, and relatively conservative growth projections suggest 8 bn in 2030, 10 bn in 2070. Such growth affects demand for food, water, space, energy and other resources; produces environmentally damaging by products and effects; depletes the stock of non-renewable resources and threatens natural processes of renewal. It tends to concentrate in urban environments, ironically contributing to rural depopulation, with concomitant social and environmental pressures, and tends to be unbalanced between rich and poor, reinforcing patterns of inequality and deprivation, threatening social and political order, destroying traditional patterns of family and community interdependence, both within and across states, cultures and continents. Such growth evokes questions relating to religious, political and personal values, of the social and moral order, of distributional equity and property rights, and the legitimacy of legal systems and processes. Apart from international effects, it may exacerbate national conflicts and threaten national boundaries. Population expansion and movement is nothing new, as the history of Indo-European and pre Colombian peoples demonstrates. Nor are the horsemen of the apocalypse unfamiliar - war, pestilence, famine and other 'acts of God' (to use the insurance term). What is new is the evidence that population expansion may be reaching the limits of Earth's capacity.

Deforestation

Wood has long been used for fuel, building and artefacts, and forests have long been cut down for pasture and agriculture. In Roman times Britain, Germany and much of Northern France were covered with forests, and 'not much has remained of the original luxuriant forests of the Mediterranean region, nor much of the soil that long ago covered the hillsides' (Hillel, 1991, p.177). The romantic rural landscapes of literary and artistic nostalgia are the products of man's husbandry. However while losses of forest have relatively stabilised in Europe (and the USA), they proceed apace in the great tropical forests, in South America, Central Africa, and South East Asia, with serious implications for local ecology and global climate. According to UN estimates, for tropical areas as a whole, 10 hectares are being cleared for every one planted (Hillel, 1991, p.179). Such forests are reservoirs of biodiversity, influence global rainfall and climate patterns, and, once lost, are irreplaceable within centuries. But they represent short-run economic resources to the countries of those regions.

Desertification

Again, this is no new phenomenon. Hillel suggests (pp.86/87) that silting and salination account for the decline of the first great urban civilisation - the Sumerians in Mesopotamia. In Roman times North Africa provided surpluses of corn for Rome and its empire, much of it from terrain which is now desert. But husbandry and ingenuity can make the desert bloom - in the Sahel food production has increased steadily over the last 20 years, according to Holdgate. But loss of subsoil (particularly in hills and mountains) and laterization (the hardening of the substrate following deforestation in tropical areas) is a less remediable problem. In 'modern' agriculture intensive fertiliser application degrades the soil structure and water tables, pesticides the biological food chain, while intensive farming on an industrial efficiency model degrades the landscape and biodiversity. Such damage is reversible (apart from the loss of unique species of biota), but this requires changes in thought, calculi and practice.

Biodiversity

'About 50% of the species believed to exist on Earth are insects and other small organisms living in the canopies of the tropical rain forests' (Holdgate, 1991, p.18), so these in particular are threatened by deforestation. In addition islands, remote valleys, mountains, deserts are specialised reservoirs of diversity and vulnerable to invasion. Biodiversity may matter in ways we do not currently understand. In particular tropical plants and other biota may (some certainly do) contain curative medicinal components, with properties familiar to indigenous populations, but compounds unknown to the corpus of modern science. And there may be unrecognised interdependencies (food chains, ecological relationships) in which an apparently minor change may undermine a whole ecosystem - the 'straw that breaks the camel's back', a phenomenon familiar to ecologists and evolutionary biologists, but now becoming more recognised as applicable in all complex adaptive systems through the study of chaos and complexity (Lewin, 1993). In short many apparently flourishing systems (including civilisation) may be balanced, indeed owe their exuberance to location, on the 'edge of chaos'. Questions here relate to kinds of resilience and adaptability within a stability domain. Forty percent of the corporations in the Fortune 500 twenty years ago no longer exist in their then form.

Ironically diversity at a high level of biological classification (phyla) has significantly reduced over time, for the Burgess shale dating from middle Cambrian times (c 530 million years BC) has revealed at least 14 distinct phyla, of which only four now survive (Gould, 1989).

Pollution

In Holdgate's judgement (p.19) 'some kinds of damaging pollution are unstoppable over a 40 year time-scale because they are already present in the environment and cannot be eliminated before the year 2030'. For instance PCBs and chlorinated organics are dispersed in ocean waters and biological food chains, and even if CFCs (responsible for depleting ozone with an 80-100 year residence time in the ozone) are eliminated from production by 2000 AD, 'the ozone hole will not begin to fill in in less than 40 years'. Likewise the build-up of greenhouse gases, notably carbon dioxide, is unlikely to be stopped by 2030. An international conference hosted by the Canadian government forecast 'effects second only to global nuclear war', if greenhouse gases are not effectively reduced (Radford, 1990, p.10). The longest lived pollutant, and most psychologically disturbing in the popular psyche, is nuclear radiation, whose avoidance, through accident or war, is dependent on industrial/scientific procedures and human control systems, and the maintenance of international order.

In addition, agricultural, industrial and economic development produce myriad other polluting by-products and dislocative effects - urbanisation and transport, domestic and industrial waste, dioxides of sulphur and nitrogen - which impact on millions of people, locally and regionally, in air, water, food and visual amenity, with psychological and social disequilibration.

Climate change

There is much professional disagreement about the likely extent of climate change and of its effects. However Holdgate's conservative summary of forecasts is that 'there is every reason to believe that by 2030 the world will be between 1° and 2° warmer than it is now', with significant regional variations, and 'wet areas may get wetter, dry areas drier and the force of storms may increase. By 2030 the sea level may rise by 10 to 20 centimetres'. The situation forty years beyond this will depend on success in curbing greenhouse gas emissions.

In sum, though there are great differences in technical extrapolation of trends, in diagnosis of pollutants and the extent of their effect, and in our limited understanding of the capacity of ecosystems to absorb and adapt, there is consensus among environmental scientists (and in related relevant disciplines) about the direction of the trends and their significance. It is this incontrovertible evidence from a wide community of professionals which provides the background to current personal and political concern for the environment. In addition, at the human level, individuals can sense their environment degrading, in air, congestion, population growth, social disorder, and so on. At the 1994 American Association for the Advancement of Science meeting in San Francisco, Professor Pimentel of Cornell University presented the dilemma as a global social choice - 'Does human society want 10-15 billion humans living in poverty and malnourishment, or 1 to 2 billion living with abundant resources and a quality environment?' (The Guardian, 22.2.94). Current population is over 5.2 bn.

Characteristic problems

In broad terms, the characteristics are simply stated. First, much pollution and damage is the aggregate result of innumerable small dispersed actions by a very large number of people and organisations. Some sources, however, are concentrated - notably sulphur dioxides from power stations, and nuclear and other civil and industrial disasters. The sources, then, are typically far removed from their effects, crossing regional, national and political boundaries, and affecting people and ecosystems in many ways. So those who suffer the impacts and bear the costs are often remote, and under different legislative domains, from the sources. The effect of carbon dioxides on climate is a classic example. Second, for many of the trends there is substantial time lag between cause and effect, and effects may be long term and lingering - notably with PCBs, CFCs, and biological and genetic damage.

These two general problems are classic cases of market failure, or, more precisely, since markets are social inventions, of the incapacity of markets to generate resolution. Indeed Adam (1993, p.407) argues that 'an environmental ethic is difficult to reconcile with economics of both the capitalist and communist kind', an echo of Means' (1988) observation that capitalists will destroy the world only for profit, Marxists will do it even without that constraint. The first characteristic requires a global value change (voluntary action) and/or a global framework of legislation, regulation and enforcement, by which real environmental costs can be signalled and enforced in the calculus of consumption and production, with mechanisms for transnational compensation (enforced action). The estimate of such costs and compensation provides rich ground for dispute. The second characteristic involves the classic problem of intergenerational choice - short term gains (or avoidance of costs) with massive costs to be borne in the future, costs which can only be known with hindsight. These two characteristic problems subsume a number of subsidiary problems, social, cognitive, political and economic:

The identification, measurement and diagnosis of environmental effects and causes requires a degree of scientific expertise, though some kinds of pollution and degradation are already evident to the layman. In 1964, Plumb noted that in the humanities specialisation 'has proliferated like a cancer'. The scientific estate has likewise become highly specialised, both between and within disciplines. But analysis of environmental phenomena requires analysis and conceptual models to match their features - a multi-disciplinary, open-minded approach, with integration across disciplinary boundaries, and less insistence on inappropriate certainty and proof of the kind demanded by sceptical inquiry.

Different disciplines use different modes of enquiry and procedures, with different foci and values, so cross-disciplinary consensus and understanding may be that much harder to achieve. Furthermore, research agendas reflect funding sources - often industry and government - so the direction and focus of research may be that much harder to shift: in energy, far more is spent on research on existing technologies than on potentially beneficial sustainable/renewable technologies. Not only are some characteristic problems difficult and expensive to research, requiring a degree of speculative interpretation (climate, the ozone layer, effects of global warming, longer term biological, ecological and genetic impacts), but the procedure and conceptualisation must fit the problem characteristics, diagnosed by Adam (1993, p.410) as 'ecologically networked interconnectedness, simultaneity and instantaneousness, out-of-sync time-frames, multiple time-lags and latency periods, non-causal connections and actions at a distance'. With long term effects, action is required early in the light of long-term forecasts, and, as Niels Bohr put it, 'prediction is very difficult, especially about the future' (Coffey, 1983).

The 20th century has seen triumphs of science and technology through industrialisation and economic growth, with measurable increases in consumption and economic (not necessarily social or psychic) welfare. This has translated into a popular quasi-religious belief in the power of science and technology to provide solutions, a belief which extends to the human sciences. But though our understanding of bodily and brain functioning, of social and institutional relationships, and of aspects of psychology, may deserve this accolade, our ability to control ourselves, our emotions, neuroses and drives has not been demonstrated - rather the reverse. As Dixon (1987) asks, in his preface to an analysis of the psychological context of civil and military disasters, 'how could a species so talented in so many ways be so incompetent when it came to ensuring its own survival?' Just as our artefacts have distanced ourselves from the external environment, so we have distanced ourselves from our inner selves, denying ancient wisdoms and eroding traditional coping mechanisms.* Close acquaintance with the world of organisations reveals failures to relate, to integrate and communicate, even for common instrumental ends. Though the external world has changed much over 3000 years, man as a sentient biological being has not. 'Men with the emotional baggage of stone-age hunter gatherers today engage in private car races on the highways, and, as leaders of the superpowers, control weapons and armies capable of igniting the world. Man's striving for power is particularly problematical since it is not turned off by biological control mechanisms' (Eibl-Eibesfeldt, 1989, p.719). In stone age times, the power over nature and over others that an individual could acquire was as limited as his technical means. In addition 'we show less concern for Man in the anonymous society in which individual interest often predominates. People are much more inclined to behave ruthlessly toward those they do not know' (Eibl-Eibesfeldt, op.cit.). Indeed the standard (and necessary) coping device in mass society (towns/cities) involves emotional distancing from the physically proximate. Civilisation (etymologically participation in city life) necessarily involves distancing and depersonalising of relationships, the weakening of personal ties and cultural sanctions. As the evidence accumulates that we are close to the limits, in our demands on our environment and the tolerance of the biosphere, a substantial consensus is emerging that science and technology, the handmaidens of the progress which has in part generated the problems, will not provide the solutions - though sustainable technologies, energy efficient artefacts, non polluting energy generation, will help. As Holdgate puts it (p.20), 'the solution to these problems, which are growing increasingly graver, does not lie in science or technology, but in politics. The lesson of the recent decades is that scientific understanding and technological ingenuity are not the limiting factors. The blockage lies in human perception and the willingness of people to change how they behave at the individual, group and national levels.'. And, as Dixon (op. cit.) wryly observes, 'politicians and heads of states should be grateful to know that they have become the most important people on earth, vital to our survival. It is because of this that the central theme of this book, i.e. that we are our own worst enemy, applies most particularly to them'.

Though much environmental damage results from innumerable dispersed actions, some sources are easily identifiable (as with Bhopal, Exxon Valdez, Chernobyl). Where source and effect lie within the curtilege of a single legal system, legal restitution may be feasible, and provide both compensation and deterrence. Punitive damages have just been awarded against Exxon of $5 bn (Guardian, 20.4.94), in addition to clean-up costs, restoration payments and private settlements. The systemic effect is that marine insurers are refusing to provide certificates of Financial Responsibility (demanded by the US Coastguard), 'for fear they will be sued for unlimited damages if one of their policy holders is involved in a spill'. Such lack of insurance could drastically affect US oil imports and prices. But most disasters and most environmental damage are not so geographically identifiable nor justiciable, so conventional legal and insurance remedies are not feasible; nor are costs and damages easily computed, even in the best case of proven liability and scientific consensus on cause and effect. In the case of identifiable potential disaster sources and pollutants, there are two problems. First, however good the design and installation of technology and safety systems, however good the training and procedures for the human operatives, the probability of high-cost disaster can never be reduced to zero (though estimates may compute it as negligible). For sociotechnical systems are not infallible, nor are the capacities of their designers (Turner, 1994). In short, concentration (large tankers, massive power generation) tends to violate the precautionary principle. Second, some risks fall outside the capacity of conventional insurance systems. With nuclear plants in particular, 'we have suspended the principle of insurance not only in the economic but also in the medical, psychological, cultural and religious sense. The residual risk society has became an uninsured society with protection paradoxically diminishing as the danger grows' (Beck, 1992, p.101). Oil spills in US coastal waters seem to be approaching the point of uninsurability. Furthermore, compensation (for loss of life or health) does not provide restitution. The problems in the Lloyds insurance market (beyond failures in regulation) seem also to indicate the increasing evidence of disasters and damage - perhaps to the point of uninsurability.

The feature which makes environmental problems particularly intractable is self-interest. Yet this is the principle enshrined in our industrial practice and our economic ideology, with rewards of power, status and privilege fuelling the quest for growth. The prevention or alleviation of damage may require great cost to be borne at the personal, firm, industry or national level, with little discernible benefit. In environmental matters personal benefit and public good are inversely related - the converse of Adam Smith's invisible hand: the environmental agenda demands collective interest, at personal cost. Can such a system be devised and enforced? In the absence of a value change, self interested agents will only be induced to change behaviour (incur costs) through compensation or preventive legislation. But those most affected by legislation are likely to express their concern over law and enforcement both vociferously and effectively, through political processes of voting, lobbying and obstruction. Exxon intend to appeal against their punitive damages, and oil importers are likely to use their lobbying and related powers to dilute or change current US legislation. And will US citizens accept a dramatic rise in the price of gasoline to cover the risks of potential oil spills in waters under US jurisdiction?

The answer to this rhetorical question, if value change is required is 'yes, they must: and that's just for starters'. In industrial societies (in contrast to agricultural and pastoral) people are so programmed to expect material progress that for many the maintenance of the status quo is unthinkable, let alone a voluntary reduction in living standards. Yet for most of their life on earth humans have maintained a steady state materially, and the majority of the world's population lives in economic poverty (not necessarily misery). The pre Colombian tribes inhabited North America for tens of thousands of years, with nothing of the environmental degradation achieved in 400 years by mainly European immigrants. So there are many models for successful survival from past history and contemporary anthropology. Such value change connotes more than the 'cultural change' of managerial rhetoric, or changes in consumption patterns. It means changes in ways of seeing, relating, thinking, believing.

Post-modernism is fashionable in academic and aesthetic circles, even extending to organisational analysis: it broadly connotes a loss of faith in old certainties and standards, with consequent eclectic experiments in style and manners. The Hellenistic period showed a similar post-classical tendency. Hellenistic philosophers (Epicureans, stoics, cynics) were, beneath their intellectual differences, united by a common theme, a search for ataraxia - peace of mind, freedom from anxiety (Grant, 1982). The source of this anxiety was external (material things, relationships) and internal (ambition, jealousy, greed and so on). It was on modes of coping that philosophers disputed. 'Nirdvanda (freedom from opposites) is the Orient's remedy for this' (Jung, 1993, p.177). In contemporary society stress related illnesses are on the increase, as is the popularity of stress management courses and of new religions. In these, and many other straws in the social and environmental wind, a future historian might see signs of a return to more spiritual, less material values (pace the Protestant ethic). But whether such values can translate into collective global self-regulation, in a short time period, is a question yet to be resolved.

Energy and environment

In terms of the above characteristics the energy industry's environmental impacts may seem relatively easy to identify and analyse. Power stations in particular are identifiable sources with identifiable outputs, a condition which helps frame solutions in terms of costs, regulatory practice, technology, fuel source and so on. Of UK pollution sources, power stations in 1988/89 generated 71% of sulphur dioxide, 32% of nitrogen oxide, 33% of carbon dioxide (quoted in Smith, 1993, p.3). Both source and emissions are concentrated, a fact that carries its own risks, in that the consequences of a disaster are magnified. From the study of surviving systems in ecology and biology two key related strategies can be discerned (Woodward, 1982) - 'maximise flexibility' and 'minimise the cost of things going wrong'. Man is biologically adaptable, to a degree, and his intelligence - generated artefacts make him highly flexible, but his social systems have generated dependent inflexibility. Concentration, with economies of scale, tends against these two principles.

Some of the papers in this volume outline the relative success in Japan in reducing some emissions: and the elimination of smog from London in the 1950s suggests a triumph of technology and law (the Clean Air Act). However, this was achieved in a context of a switch from coal for domestic use to electricity and gas (itself subsequently converted from towns to natural). But the concentration of coal-burning in power stations has exported, via high stacks and prevailing winds, sulphur dioxides as acid rain to Northern Europe. Likewise the current 'dash for gas' in electricity generation may produce concentrations of pollutant previously thought harmless in dispersion, and will accelerate depletion of gas reserves.

In short, primary energy sources - coal, oil, gas, nuclear - are embedded subsets of the ecosystem. Even apparently benign renewable technologies, whether dependent on sun, wind, wave, water or wood, have particular sets of associated environmental problems. So though the short-term specifiable characteristics of environmental damage and regulation in the energy sector make the problems and incremental agenda relatively amenable, the systemic and time interdependencies locate energy firmly in the overall environmental agenda, physically, socially, and politically. In a detailed study of the relationships between energy and civilisation through the ages, the authors (two historians and a physicist), noting the direct consequences of 'irresponsible' USSR and Eastern European resource management in Western Europe, suggest that 'the ecological consequences of energy production in China and India, while currently localised in Asia, will affect the biosphere tomorrow' (Debeir et al., 1991, p.231). Their conclusion is that 'until a massive transfer of resources allows the latter (sc. the poor) to promote a form of development compatible with the improvement of the environment, appeals for a rational approach to energy will remain a dead letter'. Their solution echoes that of others who have pondered the environment, from different disciplinary and cultural bases - 'as our air, our water, our food, the earth itself become crisis issues, the urgency of an energy revolution becomes more and more vital ... inseparable from ... a political, social and cultural revolution with no historical precedent'. Like Cobbett's observations in his Rural Rides after the English enclosures, they see urbanisation (cf. Cobbett's, 1967, description of London as 'the great wen') as a critical cause.

Such revolution would involve more than radical change in costs and prices; it concerns values, perceptions and modes of analysing and diagnosing, and institution-building and regulation.

Environmentalism: Perspectives and values

'The conceptual problem-solving apparatus that Western culture has developed is incapable not only of solving the societal problems, but even of perceiving those problems', observed von Foerster (1977, p.105), drawing lessons from biology for the analysis of complex systems. von Foerster's concern was twofold. First, the nature of societal problems is systemic, interconnected and long-term, with analysts' and policy-makers' perceptions and actions themselves part of the problem space: and second, reductionism and paradigms of causation (in a tradition which S.J. Gould dubs 'physics envy') are incapable of 'explaining' the behaviour of such systems.

Eighteen years on, the study of characteristics and patterning of complex adaptive systems is becoming increasingly popular across academic disciplines (Lewin, 1993, Waldrop, 1992), while environmentalism, which requires a multi-disciplinary holistic perspective, has engaged the attention not just of academics, but of politicians and citizens. But under the banner of environmentalism different underlying values and attitudes are discernible, with quite different implications for action. However, in different ways, it echoes many previous critiques of materialism across time and culture, and incorporates attitudes discernible in less developed or 'primitive' societies - primitive materially perhaps, but not biologically or socially. The following classification uses terminology coined by Tim O'Riordan (1995), but, like all such classifications of social phenomena, provides approximate typology rather than inclusive definition - clustering on a multidimensional spectrum.

First, 'dry greens' see environmental effects as being incorporated by fiat into the economic calculus, signalling costs and priorities via market mechanisms, and mediated through legislative, fiscal and regulatory intervention. With significant implications for property rights and valuations, these interventions would be politically legitimised by democratic processes, and/or entailed by international conventions. In broad terms this is the approach taken by 'environmental economics' (e.g. Pearce et al., 1989; Helm, 1991) and articulated in some of the papers in this volume. It is characterised by a form of conservative incrementalism, starting from the present, with a relatively short time-frame, maintaining and working through existing political and economic structures, but with radical implications for costs, prices, and behaviour. In short, same game, same players, different scoring in terms of prices and costs.

Second, 'shallow greens' see mankind as dependent on the adaptive capability of nature, a 'service' which must be appropriately costed in economic calculi. But in the light of environmental trends and their characteristics, strong emphasis is laid on the principle of precaution. Thus it rests with a proposer to demonstrate that environmentally harmful effects of a development can be accommodated: where doubt exists, precaution must be the guiding principle, with buffering allowed for current ignorance. In this formulation, industry is 'licensed' to practise on condition that it preserves the life-saving functions of the natural world. The implications of this view, elements of which are discernible in national and international law (perhaps a result of disasters and accumulating evidence) are profound for incremental decisions; if applied retrospectively, more profound. In sum, same players, new kind of game.

The above two positions are characterised by their calculative approach to nature. The third, 'deep green', echoes elements of religious beliefs common to early mankind and some contemporary mythologies (Campbell, 1964), which, through a variety of myths of origin and purpose, see mankind as part of the great chain of being, the natural mysteries of birth, life, death and renewal. Life has intrinsic mystical value, and the role of humans, as children of earth, is trusteeship and propitiation. This position takes many forms, from withdrawal, as in the formation of utopian communities (early Cistercian communities, various contemporary 'back to nature' movements), through right living (the green citizen, rather than the green consumer) to engaging in public political debate and action. All emphasise sustainability, with a tendency to moderation and self-sufficiency (needs rather than wants). The time-frame also tends to be long, echoing a sense of deep time and the relatively recent place of mankind in earth's evolution, which is discernible in scientific writings on geology, palaeontology and evolutionary biology (7000 years of 'civilisation', 100,000 years of homo sapiens, 600 million years of multicellular life, 5 billion years of planet earth). In short, radically different game, different players, different calculi.

The spirit is well captured in an eloquent, perhaps prophetic, speech of Chief Seattle in 1854 (quoted in Mitroff and Linstone, 1993): 'How can you buy or sell the sky, the warmth of the land? ... one portion of this land is the same to him (sc. the white man) as the next, for he is a stranger who comes in the night and takes from the land whatever he needs. The earth is not his brother, but his enemy, and when he has conquered it, he moves on .... Whatever befalls the earth befalls the sons of earth. Man did not weave the web of life, he is merely a strand in it. Whatever he does to the web, he does to himself ... the whites too shall pass, perhaps sooner than all other tribes'.

Attempts have also been made to locate 'green' viewpoints/values towards one end of a spectrum of scales between opposites, the other extreme being 'free market industrialisation' (Porritt, 1984) or 'technocracy' (Lorenz, 1984) - on economy (growth versus sustainability; production for exchange and profit, versus production for use and need); on politics (centralisation versus decentralisation; representation versus participation; aggressive individualism versus co-operation); on organisation (hierarchy versus communities and networks); on environment (domination over versus harmony with nature; resource exploitation versus husbandry): on time (short term versus long term); on thinking and feeling (rationality versus intuition; technocratic knowledge versus empathy/understanding). But these opposites are not clear cut, and just as they reflect tensions in societies, also reflect tensions in individuals: in society, resolved (ideally) by political negotiation, bargaining, compromise and accommodation, in individuals, by psychological mechanisms of denial, attribution and projection, or hopefully of psychic integration. But consistency, of rhetoric or behaviour, is not a characteristic of political systems, of consumer or religious behaviour, or of rational mankind.

But however one diagnoses the differences within the spectrum of green views, the implications are profound, ranging from action under compulsion - significant and increasing regulatory intervention, nationally and internationally - to voluntary action, a global value shift reflected in living styles and standards, or a longer term Malthusian acceptance of the 'natural retaliation of Mother Earth'. In the absence of a profound global value shift, what forms are such interventions likely to take, if the accumulating evidence is valid, and environmentalism more than a moral panic, a passing millenial fad?

If the solution lies in politics, then there is an emerging consensus on the form this will need to take - international legislation, which must in turn be policed and enforced. Such legislation would require agreements which curtail national sovereignties and individual choices, entailing problems of subsidiarity far greater than those visible as the European Union attempts common policies. They would also - to be universally accepted - require massive transfers of resources from the developed to the developing world, to compensate for maintenance of national resources (forests, land, ecologies and so on) and for non-polluting technologies. For from a third world perspective, the industrialised developed nations would be asking the less developed to forego the industrialisation which has brought the material prosperity of development, and which has itself generated the environmental problems. But the history of international agencies has an uneven record, characterised by bureaucratic growth, goal displacement, and problems of enforcement and compliance of proclaimed agenda. Furthermore an apparatus of law and hierarchy, requiring prevention rather than promotion, goes against many 'green' notions of voluntaristic co-operation and community. Such international consensus would have to rest on political foundations, requiring a degree of political acquiescence from all nations which would itself constitute a significant value shift. Such a change is only likely to occur as environmental degradation and climate change impacts further on life-styles, in local disasters and global change, with accompanying breakdown of communities and international order.

Conclusion

I have attempted to summarise the environmental context for energy regulation by drawing on a variety of sources for evidence of environmental trends, but have not done justice to the complexity of the scientific interrelationships, nor the extent and variety of the evidence. But two points emerge with great clarity: first, the complex long-term interdependence of these trends and second, the consensus on their direction and the need for political resolution. Differences related to the interpretation of evidence, the extent and timing of their significance, and appropriate and/or feasible behavioural, institutional and political responses, for 'appropriate' and 'feasible' often appeared diametrically opposed. The case of energy - its pollutants, sources, technologies, distribution - is part of the much wider problem. I have used 'energy' in its restricted industrial sense, but in ecology both living and inanimate systems are diagnosed as energy users, converters and stores. Indeed, for environmental accounting Radford (1990) suggests that an energy-based numeraire (reflecting environmental value) would be more appropriate than a monetary numeraire (reflecting scarcity and human valuations). What would that mean for accounting, economics and finance?

The tenor of this chapter certainly points to a difficult, to put it mildly, scenario, and it would seem appropriate, in an academic dialectic tradition, to put forward an opposite scenario which provides grounds for optimism, or, at least, continuance of the status quo, with incremental, though substantial, accommodation of technical and regulatory mechanisms for reducing pollution. But I have been unable to find evidence or argument to set against the accumulating evidence provided by those whose business it is to monitor and interpret these trends - botanists, biologists, ecologists, physicists, chemists, demographers, psychologists, indeed all those scientific disciplines which contribute to 'environmental science'. Counter arguments tend to take the line that radical remedial action is personally, economically, politically, or organisationally infeasible (Cairncross, 1991). And of course they are right, for they relate to the 'real', though socially constructed, world of economic and political interdependence. Unlike the real world of liberal economic ideology, relevant actions tend not to generate individual pay-offs (though there may be competitive market and strategic advantage to firms cultivating a 'green' image, or in investing in green technologies and practices above or ahead of regulatory requirements): rather they impose great identifiable personal, corporate and social costs for a dispersed collective global pay-off. In short public goods dominate private goods. This is the burden of the precautionary principle which seems likely to be increasingly enacted and enforced.

Contemporary business literature (if it may be so called) is replete with cases of firms where the internal reality of organisational values, perceptions and processes is far removed from the external market realities (Woodward, 1982, and Mitroff and Linstone, 1993, whose first chapter is entitled 'The World That Was and Is No More'). The solution is painful and radical internal change - downsizing, re engineering, cultural change, quality, empowerment, and a host of other fashionable prescriptions. Often the external 'reality' is denied by those in authority, who have retired with pensions or golden handshakes by the time their failures of responsibility are manifest. The outcome is either painful internal adjustment, take-over or bankruptcy. For liberal economics liquidation of inefficient firms through competition and take-over is a sign of healthy economic activity. But for a healthy environment, what would liquidation of environmentally inefficient firms, households, institutions and countries entail? The key point is that counter arguments to evidence of environmental trends tend to deny or ignore the evidence, and there are strong parallels to this psychological tactic in contemporary corporations and in other contexts. Dixon (1987) has provided ample documentation of cases of military and civil disasters where key actors would rather die (along with those dependent on them) than violate their internally constructed peace of mind, through repression and denial. And of course, for peace of mind, one can point to a plethora of forecasts and doomsayers who have been notoriously wrong in the past.

At the start of this chapter the question was posed whether current concern for the environment might seem quaint in fifty years time. My guess is that it will: not because it is wrong (the common air is free and in infinite supply) or socially and psychologically odd (the place of women in business), but because the warning signs were not sufficiently acknowledged.

However, to conclude on an optimistic note, though we may be approaching the limits of growth and material progress, there will surely be growth and progress in studies of regulation. To this end the contributions to this volume constitute a step in the right direction, though hindsight may suggest that they are only a beginning.

References

Adam, B., 1993, 'Time and Environmental Crisis: an Exploration with Special Reference to Pollution', *Innovation*, 6:4, pp.399-413.

Beck, U., 1992, 'From Industrial Society to the Risk Society: Questions of Survival, Social Structure, and Ecological Enlightenment, *Theory, Culture and Society*, 9, pp.97-123.

Cairncross, F., 1991, *Costing The Earth*, Business Books.

Campbell, J., 1964, *The Masks of God* (3 volumes), Penguin.

Chapman, S., 1935, *Modern Business: Principles and Practice*, Odhams.

Cobbett, W., 1967 *Rural Rides*, Penguin (first published, 1830).

Coffey, W., 1983, *303 of the World's Worst Predictions*, Tribeca.

Debeir, J.C., Deléage, J.P. and Hémery, D., 1991, *In the Servitude of Power*, Zed Books.

Dixon, N., 1987, *Our Own Worst Enemy*, Futura.

Eibl-Eibesfeldt, I., 1989, *Human Ethnology*, Aldine de Gruyter.

Gould, S.J., 1989, *Wonderful Life*, Penguin.

Grant, M., 1982, *The Hellenistic Greeks*, Weidenfeld.

Helm, D., 1991, ed., *Economic Policy Towards the Environment*, Blackwell.

Hillel, D., 1992, *Out of the Earth: Civilisation and The Life of the Soil*, Aurum Press.

Holdgate, M., 1991, *Environment*, 33:6, July/August.

Jung, C.G., 1993, *Memories, Dreams, Reflections*, Fontana.

Lewin, R., 1993, *Complexity*, Phoenix.

Lorenz, K., 1984, *The Waning of Humaneness*, Penguin.

Lovelock, J., 1990, *The Ages of Gaia*, Bantam.

Means, R., 1988, 'Fighting Words on the Future of the Earth' in Zerzan, J. and Carnes, A., eds., *Questioning Technology*, Freedom Press.

Mitroff, I. and Linstone, H., 1993, *The Unbounded Mind*, Oxford.

O'Riordan, T., 1995, ed. *Environmental Science for Environmental Management*, Longmans.

Pearce, D., Markandya, A. and Barbier, E.B., 1989, *Blueprint for a Green Economy*, Earthscan.

Plumb, J.H., 1964, *Crisis in the Humanities*, Pelican.

Porritt, J., 1984, *Seeing Green*, Blackwell.

Radford, T., 1990, *The Crisis of Life on Earth*, Thorsons.

Smith, D., ed., 1993, *Business and the Environment*, Paul Chapman.

Turner, B.A., 1994, 'Causes of Disaster: Sloppy Management', *British Journal of Management*, Sept.

von Foerster, H., 1977, 'The Curious Behaviour of Complex Systems: Lessons from Biology', in Linstone H. and Simmonds, W.H., eds., *Futures Research*, Addison-Wesley.

Waldrop, M.M., 1992, *Complexity*, Viking.

Woodward, N., 1982, 'The Myth of Turbulence', *Futures*, August.

* Note: The rhetorical shift to the first person plural is deliberate, inviting a degree of self-reflection.

PART 2

Assessing Regulatory Impacts

CHAPTER 2

Analysis of the effects of environmental regulation: A case study on countermeasures against sulphur dioxide in Japan

Akihiro Kuroki

Introduction

The Japanese approach towards controlling sulphur dioxide emissions may serve as a model of success in dealing with pollution problems. For Japanese policy measures against sulphur oxides, carbon monoxide, and photo-chemical smog have been largely successful in reducing these pollutants. In particular, measures against sulphur dioxides have proved extremely effective. However, the investment required to implement these measures has amounted to more than half the entire plant and equipment investment in a given sector at the time, and the resultant impact on the economy as a whole has been significant. Accordingly, we have chosen this case because it demonstrates both the potential and the costs of these policy measures. However, as the actual sulphur dioxide regulations are complicated, their content has not been considered here in detail. (We will only review the approach and its outcome). In addition, there may be other contextual factors which have contributed to the successful reduction in sulphur dioxide emissions. So this paper provides interim results.

Japanese air pollution control measures

The legal side of Japanese air pollution control derives from the 1968 Air Pollution Control Act.

In 1962, a regulation on the degree of concentration of sulphur dioxide at the mouth of a smokestack was introduced as a first measure. The so-called 'K-value regulation' defined the acceptable degree of concentration of sulphur dioxide in smoke calculated according to the height of the smokestack and the location of the origin of the emission.

This regulation was gradually strengthened by including more facilities and areas as well as lowering the acceptable degree of concentration (Fig 1).

Figure 1

K-value (Tokyo Area)

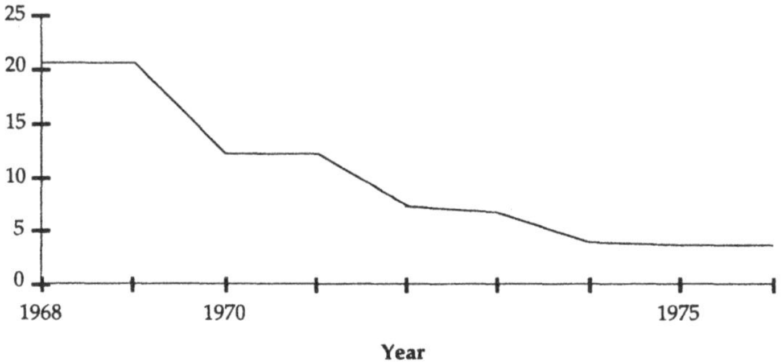

$$Q = K \times 10^3 He^2$$

Q : Maximum acceptable emission quantity

K : special value

He : Height of outlet of smokestack (adjusted by special formula)

Source: Environmental Agency of Japan 1991

However, this approach did not result in an acceptable reduction in metropolitan areas, so the 'Aerated Total Pollutant Load Control Regulation' was introduced in three areas which suffered from the worst pollution. This regulation defined the acceptable degree of sulphur dioxide concentration for each factory regardless of the height of the outlet, so that the total amount of sulphur dioxide emissions could be controlled to a given level. Of course, this created a dual set of regulations alongside the 'k-value regulation'. Table 1 outlines the series of regulations through time.

Table 1
History of Sulphur Dioxide Pollution Control Regulation Enforcement in Japan

Year	
1962	Smoke and Soot Regulation Law passed
1968	Air Pollution Control Law legislated First K-value regulation (k=20.4 ~ 29.3, 21 areas)
1970	Second K-value regulation (K=11.7 ~ 26.3, 35 areas)
1971	Third K-value regulation (51 areas)
1972	Fourth K-value regulation (K=7.01 ~ 22.2, 70 areas
1973	Fifth K-value regulation (K=6.42 ~ 22.2)
1974	Sixth K-value regulation (K=3.5 ~ 17.5, 99 areas) Areawide Total Pollutant Load Control introduced Fuel use regulation introduced
1975	Seventh K-value regulation (K=3.0 ~ 17.5)
1976	Eighth K-value regulation (120 areas)

The impact of the regulations

The change in the levels of sulphur dioxide concentration in the air in Japan clearly demonstrates the effect of these regulations through time (Fig 2). The degree of sulphur dioxide concentration was measured by averaging the figures from 15 designated locations in metropolitan areas. Following the highest figure, which was recorded in 1966, a drastic decline was maintained until 1975. Even since 1975, this tendency to decline has continued, although of a lesser magnitude. The Japanese environmental standard for sulphur dioxide is set at a daily average of 0.04 ppm. We now have only five or six locations out of more than 1,600 designated locations throughout Japan which record higher concentrations than this standard. Furthermore, all of these locations are in an area where the Sakurajima volcano has a strong influence. Thus, it can be stated with some confidence that the sulphur dioxide problem has been solved in Japan.

Figure 2
Annual average of SO_2 concentration in Japan

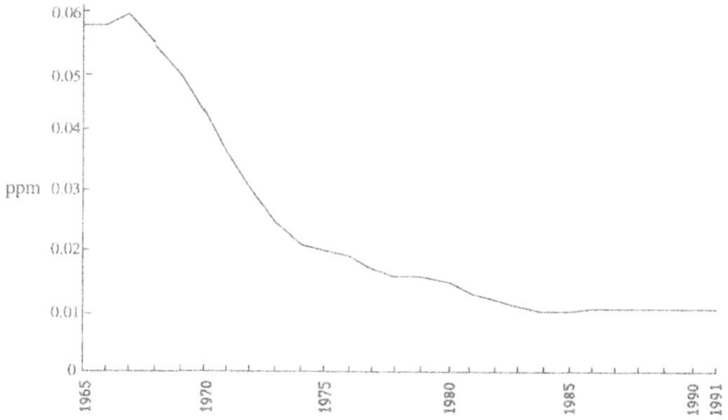

Source: Environmental Planning Agency 1993

Figure 3
Total SO_2 emission from factories in Japan

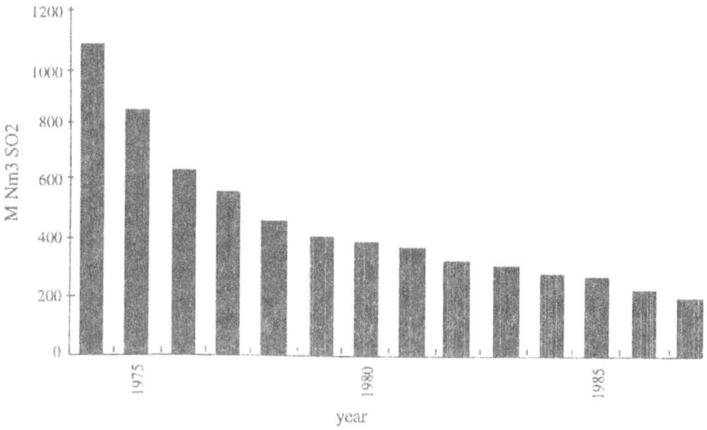

Source: Environmental Planning Agency 1988

This decline in emissions can also be seen in the statistics for the total amount of emissions from factories (Fig 3).

Figure 4 shows the rate of decline of sulphur dioxide concentrations compared to other nations.

Figure 4
Comparison of total SO_2 emission in 1975 and 1987

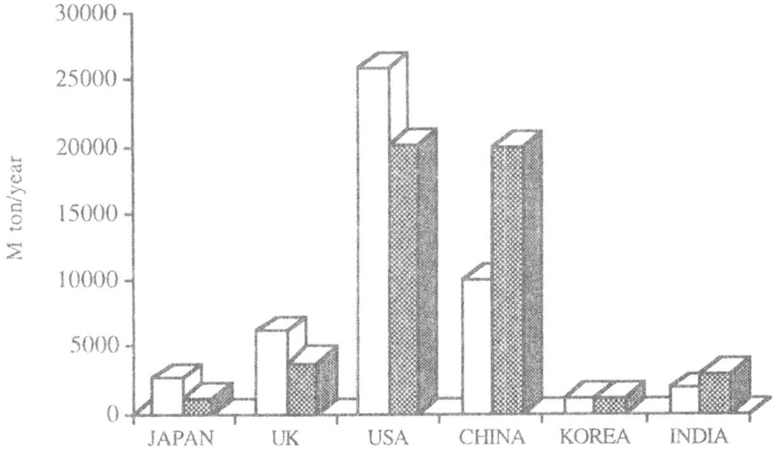

Figure 5
Total Pollution Prevention Investments in Japan

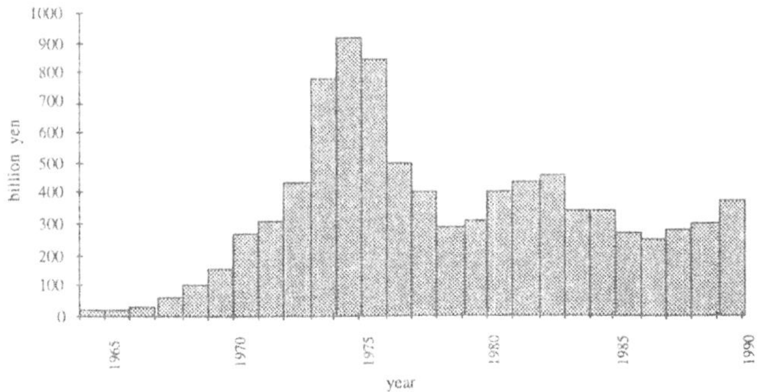

Source: National Institute for Research Advancement 1993

Effect of the regulations on the economy

Analysis of how Japanese environmental regulations have affected economic activity is limited. This may be due to the fact that public opinion has tended to support the notion that economic cost should be disregarded in order to avoid serious health hazards. In effect, the benefits of pollution reduction in reducing damage to health have been implicitly valued as outweighing any costs in terms of reduction in economic activity (or alternatively corresponding price changes).

Figure 6
Ratio of pollution prevention investment to total investment

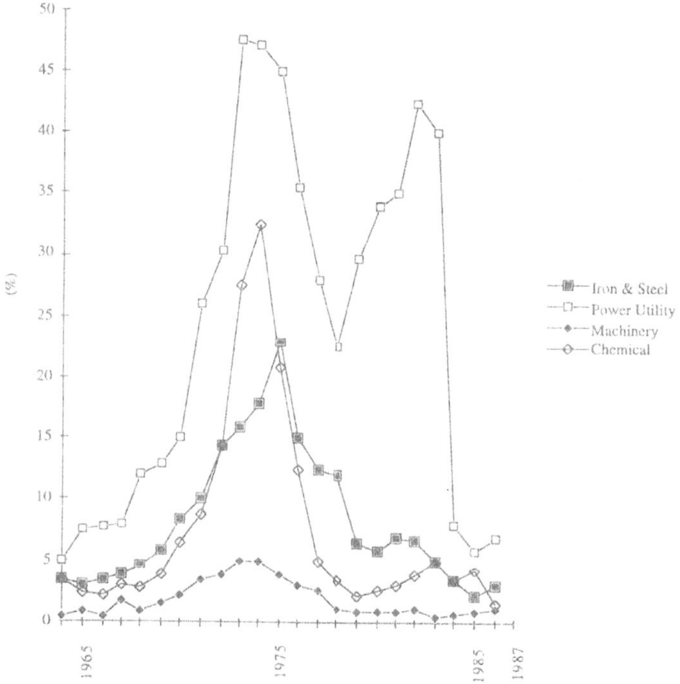

The trend in investment in anti-pollution devices in Japan has shown a rapid increase since the latter half of the 1960s (Fig 5). At one period pollution prevention investment totalled more than 17% of total annual capital spending in Japan, which indicates the extent to which environmental regulation affected the Japanese economy. Further, such investments were not evenly distributed across industries. The machinery industry carried a relatively light burden while the electricity supply industry, iron, steel and chemicals industries carried very heavy burdens (Fig 6). The investment in stack gas desulphurisation facilities specifically shows the direct relationship between regulation and anti-pollution investment (Fig 7).

Figure 7

Total investment in stack gas desulfurization facilities in Japan

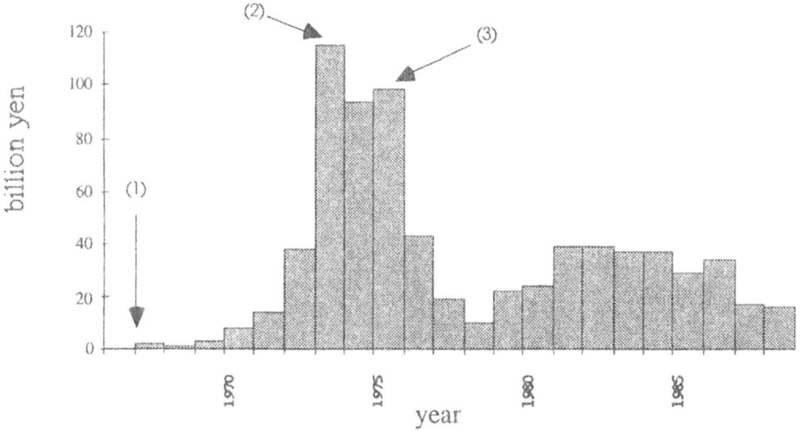

(1) First K-value regulation starts
(2) Area-wide Air Pollution Load Control regulation
(3) Last (8th) k-value regulation enforcement

Source: National Institute for Research Advancement 1993

During the early period when the k-value was set high, companies met the regulation standard by using low sulphur oil. But when the k-value fell and the area-wide air pollutant load control regulation was introduced, the quantity of investment increased dramatically.

This graph also shows that the amount of investment dropped at one stage and then drastically increased during the period from 1980 to 1987. The reason for this increase will be explained later. The next paragraphs illustrate how difficult it is to forecast the economic impact of regulation.

In 1975, the Ministry of International Trade and Industry of Japan tried to estimate the impact on generation costs of anti-pollution controls.

Figure 8 shows the estimates of impact on generation cost of desulphurisation.

Figure 8

Ratio of increase of power generation cost to reduce SOx

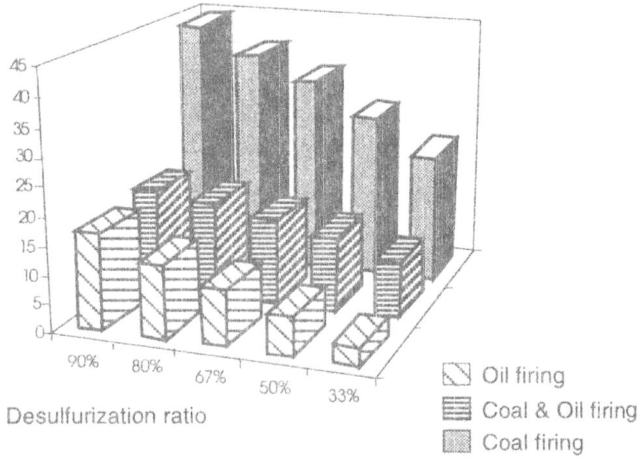

Source: Ministry of International Trade and Industry 1975

According to my preliminary estimation, the share of environmental management cost of Japanese Electric Power Companies in 1992 was roughly 14.5% of the whole cost of generation. Since the air pollution control investment contributes roughly 70% of the total environmental investment of these companies, it can be assumed that the share of air pollution cost is around 10%, which is almost one fifth of that of MITI's estimate of twenty years ago.

Measures other than direct control

Japan has used other measures in addition to direct controls, including economic incentives such as subsidies and low-interest loans. In addition, the institution of the 'Pollution-related Health Damage Compensation Law' provided an economic incentive for comapnies to decrease sulphur dioxide emission.

Figure 9

Patients recognized by Pollution-related Health Damage Compensation Law

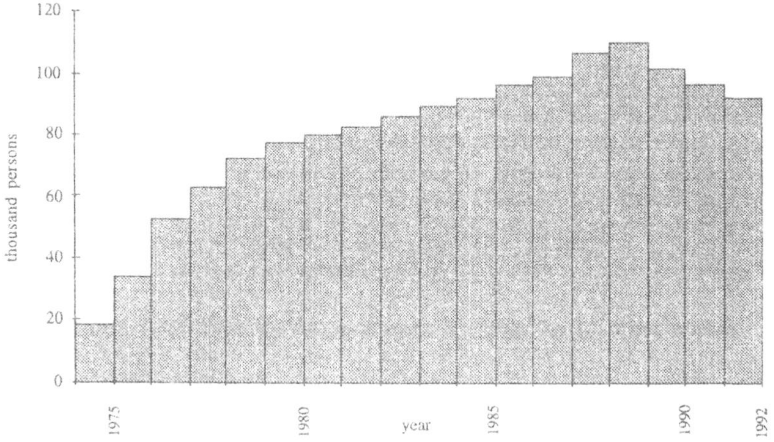

Source: Environmental Agency of Japan 1993

Figure 10

Ratio of reduction of SOx emission

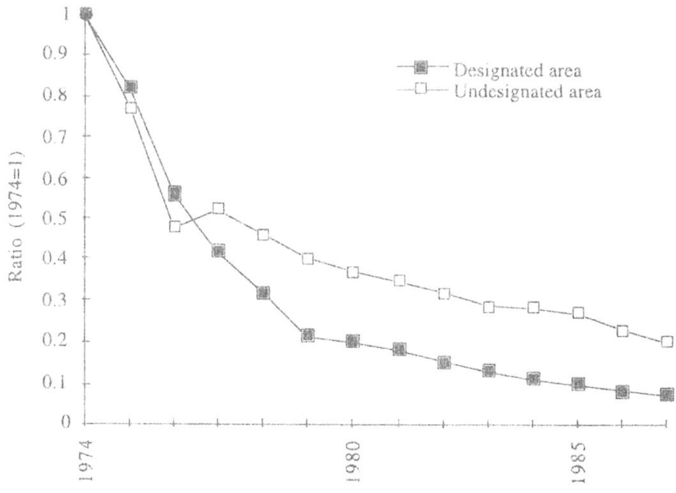

It is difficult to prove the relationship between individual illnesses, such as bronchial asthma, and the emission of pollutants by a specific firm. Therefore, the medical bills and livelihood assistance for affected patients in designated areas having intense pollution, was charged to the businesses in proportion to the amount of sulphur dioxide emitted by each firm. Since this system started in 1974, the number of patients claiming this benefit have rapidly increased, so the dues that must be paid per unit of exhaust has grown considerably. Figure 9 shows the number of patients per year recognised under the law: medical and assistance bills will have grown more steeply. Consequently, this cost provides an additional incentive to further decrease sulphur dioxide emissions. This explains why the decrease in sulphur dioxide exhaust in designated areas with high dues was greater than those in areas with relatively lower dues (Fig 10).

Analysis of effect of decreasing sulphur dioxide

The Environmental White Paper (1991) published by the Environmental Agency in Japan has given the following reasons for the decrease of sulphur dioxide emissions:

(1) Strengthening of regulations: tighter regulation standards and introduction of the 'aggregate amount' regulations.

(2) Changes in the energy situation: energy conservation and introduction of cleaner energy sources (i.e. nuclear power, natural gas).

(3) Changes in the structure of industry: energy consuming and air polluting industries are no longer central to the Japanese industrial structure.

(4) Incentive effects: the introduction of pollution victims health funds created an additional incentive to reduce sulphur dioxide emissions because of the increasing cost of the dues.

The Environmental Agency has calculated the extent to which each of these factors contributed to the results. However, their calculations are problematic for the following reasons:

(1) They ignored the economic impact of the Pollution-related Health Damage Compensation Law, though they claim that they did attempt to calculate this. However, this impact was significant, as shown by the differences in the rate of decrease between areas (Fig 10).

(2) They ignored the individual regulations adopted by local governments. Prefectures experiencing intense pollution adopted

stricter standards than those of the national government through local ordinances. These included so-called 'top-loading' regulations which require lower ceilings than national ones, and so-called 'side-loading' regulations which extend the limits on facilities included in the regulations. The effect caused by the difference between local ordinances and national regulations cannot be disregarded, since the controlled amount can sometimes be as low as one half of the national figure, depending on the kind of facilities.

(3) They ignored the impact of 'administrative guidance' by local governments. In many cases, cities, towns, and villages asked individual businesses to reduce the levels of exhaust even further, and the effect produced by such activities cannot be disregarded. According to questionnaires sent to medium-sized and small businesses regarding pollution prevention, their motivation for making investments in anti-pollution devices included tighter regulations, and guidance and advice from public authorities. These responses are summarised in Figure 11.

Figure 11

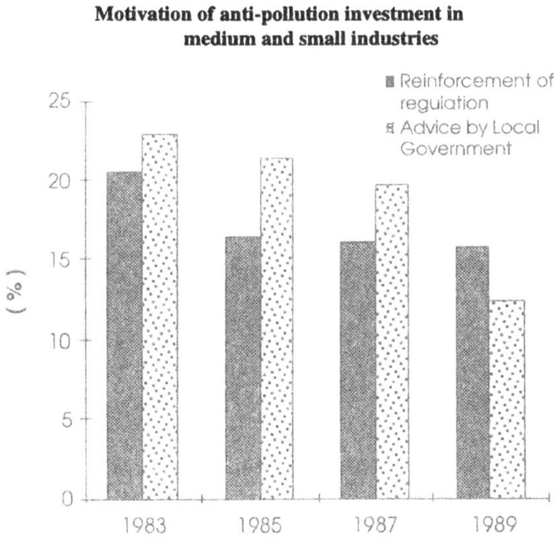

Source: Environmental Planning Agency 1992

This 'administrative guidance' was given by local governments because national governments do not provide guidance or advice to individual businesses. The impact of administrative guidance, which is not based on specific laws and regulations, cannot be ignored. Therefore, it can not be definitively stated whether or not the decrease in sulphur dioxide emission beyond the required standard was due to the incentives provided by the Pollution-related Health Damage Compensation Law or through pressure from local government regulations and administrative guidance.

Conclusion and lessons for global warming regulations

The following summarises the conclusions of our project so far, and touches on the measures against global warming.

(1) It is difficult to forecast the effect of an environmental control regulation prior to implementation. Even Japanese sulphur dioxide regulation, which is supposed to be a great success, experienced difficulties. For example, repeated tightening of standards was not effective, and a focus on the aggregated amount had to be introduced. Nitrogen dioxide control has not been as effective as initially forecast, and demonstrates the difficulty of accurately forecasting the effectiveness of such measures.

(2) It is also difficult to accurately forecast the impact of environmental control regulations. As outlined earlier, the costs borne by industry can be enormous. Indeed, the Japanese aluminium industry lost its international competitiveness due to the increase in electricity costs resulting from pollution control, and is now practically extinct. However, technological innovation triggered by tighter control or changes in the industrial environment can easily alter the magnitude of the cost, so it is difficult to make accurate forecasts. MITI's calculations were based on the information available at the time: viewed ex post it was an overestimate.

(3) Both (1) and (2) prove the importance of the gradual introduction of regulations, adjusting them according to the effect created. It is desirable to introduce regulations gradually, and not all at once, from the viewpoint of effectiveness and cost. By introducing regulations gradually, businesses have a wider variety of choices in dealing with them, and have time to adjust their plans and processes.

Finally, what are the implications of these experiences for measures to deal with global warming problems? I will review the major counter arguments to taking early action against global warming.

(1) The cause and effect relationship and the magnitude of potential damage have not been fully estimated. It is true that there are many uncertain factors involved in the relationship between the density of carbon dioxide and temperature increase or calculating the damage expected from rising temperatures. However, such arguments also accompanied the delay in response to sulphur dioxide issues in Japan. When Japanese sulphur dioxide regulations were introduced, scientific data available at the time was insufficient to determine the acceptable level of concentration in the environment, so regulations began only with exhaust control.

The lesson from this experience is that it is possible to start with temporary regulations to deal with global warming problems while fully acknowledging that scientific data is not sufficient at this point in time.

(2) It is difficult to evaluate the effectiveness of regulations and their economic impact. The effectiveness of carbon dioxide emission control and its economic impact has been analysed by a number of analysts, with inconsistent results. Nevertheless, our past experience demonstrates that gradual introduction of regulations, in spite of insufficient data, have proved to be effective to a certain extent, especially with later repeated adjustment.

(3) We have no technology to remove carbon dioxide from the environment, which is a completely different situation compared with other substances we dealt with in the past. Twenty years ago, desulphurisation technologies were not economically accessible to medium and small businesses, so they dealt with the situation through energy conservation and switching to fuel oil, which was lower in sulphur content. Reduction of carbon dioxide can be approached in the same manner, by energy conservation and altering the fuel to natural gas which emits less carbon dioxide.

In conclusion I believe that it is necessary to embark immediately on emission control of carbon dioxide as far as possible within the means currently possible, and that lack of precise data on the economic and physical effects of such an initiative should not be used as an argument of policy delay.

CHAPTER 3

Setting the objectives for environmental regulation

Nick Eyre

Defining objectives: What is regulation for?

Environmental regulation is obviously defined as regulation undertaken to protect environmental standards, and this suffices as a broad definition of its scope. But the precise objectives of environmental regulation are frequently far less clear, in some cases even to the regulator. For historical reasons, regulatory practice often is largely a process of negotiation between polluter and regulator, rather than a system based on objective criteria. Without a clear statement of objectives, it is not possible to evaluate the efficiency or effectiveness of various regulatory tools. Definition of precise objectives of the regulation is therefore the first step in creating ideal environmental regulation.

A huge variety of principles or approaches have been proposed and used as the basis for objectives setting, including:

- As low as reasonably achievable (ALARA)
- Best practicable environmental option (BPEO)
- Best available technology not entailing excessive cost (BATNEEC)
- Environmental Quality Standards (EQS)
- Integrated pollution control (IPC)

- The precautionary principle
- The polluter pays principle, and
- Sustainable development.

Out of this plethora of different, and often conflicting, principles it is possible to identify four broad approaches to setting objectives for environmental regulation. These are as follows:

Acceptable costs

This is the traditional approach in many countries, where competitiveness of polluting industries has been a priority. Where environmental and health impacts exist and the responsible polluters are identified, the appropriate regulatory agency sets environmental standards for plant emissions, primarily with reference to the costs which will be incurred by the polluter in meeting those standards.

Adequate monitoring (continuous or spot checks) is required to ensure that the standards are respected, but detailed quantitative monitoring is not necessary.

The principle used for setting standards is best available technology not entailing excessive cost (BATNEEC). Whilst this is widely acceptable as a form of words, it is not in itself a clear definition. In particular, the word 'excessive' is open to a variety of interpretations. In practice, application of BATNEEC tends to involve negotiation between polluting industries and their regulators. The results depend upon general perceptions of relative importance of the pollution and increasing abatement costs, as well as the relative power of polluting and environmental interests.

The same issues apply in long range transboundary pollution negotiations, except that there is usually a larger number of actors. The most famous example in the energy sector is the negotiation of the European Community's Large Combustion Plant Directive (Johnson and Corcelle, 1989). Designed principally to address the issue of European ecosystems sensitive to acid emissions from power stations, refineries and large industrial boilers, the proposal from the Commission included both standards for emissions from new plant and national limits on emissions from existing plant. The negotiating process took five years in the face of conflicting national, industrial, political and consumer interests before finally emerging as a directive in 1988 (CEC, 1988).

It is now clear that the objective of protecting sensitive ecosystems will not be achieved by the Directive. Negotiations under the auspices of UNECE are likely to result in revision of the Directive in due course, placing more emphasis on higher environmental standards. In this case, additional scientific knowledge and higher public consciousness of environmental issues has moved the balance of the argument towards tougher regulation.

Adequate safety and standards

This approach emphasises the impacts on the environment and human health. The underpinning assumptions are that there are 'safe levels' of pollution or levels which result in an 'acceptable risk'. The role of environmental regulation is then to ensure that these levels are not exceeded. This is achieved by the setting of standards or goals for environmental conditions, environmental quality standards (EQS). These may then be converted into emissions standards by use of suitable models. Good scientific data and analysis is required in the first instance to ensure that these standards achieve the desired objective. Adequate plant monitoring is required subsequently only to ensure that the standards are respected, but quantitative monitoring is not necessary. Improved understanding of environmental science may also result in revision of standards.

It has also been recognised that many environmental control technologies only transfer hazardous substances from one medium (air, water or land) to another. In addition, some technologies abate one pollutant at the expense of producing another. It is therefore necessary to control emissions to all media and to consider pollution control technologies in the light of all their emissions, hence the concept of integrated pollution control (IPC).

The philosophical basis of the standards is often not clear. For example, the standards operated in the nuclear industry for exposure to ionising radiation have historically differentiated between workers in the industry and the general public. The underlying assumption is that people who have chosen to work with radiation have accepted a higher level of risk of acquiring cancer than is appropriate for a non-voluntary risk. There is some evidence that public opinion would support such an assessment. But the basis of the acceptable levels of risk in each case is rarely made explicit. The definition of safe or acceptable levels has again, in practice, usually been achieved by a political bargaining process around the concept of 'as low as reasonably achievable (ALARA)', which clearly relies heavily on consideration of abatement costs.

Uncertainty poses a major problem for the setting of acceptable standards. Whilst it is frequently possible to define the effects of large (unacceptable) pollution levels, there is often considerable scientific uncertainty and disagreement about the risks resulting from smaller pollution levels. Under these conditions, epidemiological and ecological data sets are usually 'noisy' and therefore preclude any accurate assessment of dose/response functions. In these circumstances, more environmentally concerned commentators tend to urge the application of the precautionary principle, in other words assessment based on a worst case analysis.

Implicit in the 'standards approach' to environmental regulation is that it is the harm done to the most affected person (or ecosystem), which is the criterion for acceptability. There is usually little or no consideration of the numbers affected or the aggregate effect over the whole population. This approach to regulation tends therefore to impinge most strongly upon environmental impacts which are local in character, with much less severe implications for long range pollutants with a small effect on a large population. This is consistent with perceptions of equity - that a polluter should not be able to impose a severe impact on any individual. However, it bears no necessary relationship to considerations of economic efficiency. Very large investments in pollution investment may be required to produce benefits for a very small affected population. Conversely, small marginal abatement costs may not be required even where they would give significant benefits aggregated over a large population. In either case, utility (in the neo-classical economic sense) is not maximised.

Economic efficiency

This approach derives directly from the paradigms of neo-classical economics. It is assumed that the impacts in terms of environmental and health effects can be described in terms of changes in welfare of individuals affected, and that these welfare changes can be measured in monetary units. In this case, the analysis of pollution abatement is amenable to conventional cost benefit analysis. Environmental regulation is not determined primarily with respect to either abatement costs or environmental impacts (as in the previous two approaches), but with an even-handed consideration of the two. The aim is to reach the 'optimum' pollution level, at which the marginal costs of abatement are equal to the marginal costs of damage.

Cost-benefit analysis of environmental regulation has grown in popularity over recent years along with the use of market mechanisms, such as taxes and tradable permits for environmental regulation. The two are not necessarily linked, as market mechanisms can be used to meet standards based on a 'safe limit' approach, just as regulatory standards can be used within a cost benefit framework. However, both are consistent with the consideration of environmental impacts as an extension of the market economy, the dominant paradigm of economic development in most of the world. The use of market mechanisms is also consistent with the 'polluter pays' principle. This principle has gained very considerable support in recent years as it combines the theoretical economic efficiency advantages of market based instruments with the popular attraction of environmental justice.

There are practical problems in implementing this approach, which are discussed further in later sections of this paper. In general, the pollution abatement costs can be ascertained fairly accurately by engineering approaches, but the values of environmental damages are much less certain. Uncertainties in environmental modelling and valuation multiply to make most estimates of damage values no better than an order of magnitude. Any attempt to justify fiscal instruments with reference to exact estimates of damage values should therefore be treated with considerable caution. However, in many cases, order of magnitude estimates could be useful in defining priorities, developing policy and regulatory goals and identifying appropriate technologies. For example, even the rather imperfect knowledge we have of the effects of sulphur dioxide emissions is adequate to conclude that the benefits of flue gas desulphurization on new (coal and oil fired) power stations outweighs the costs.

Sustainable development

This approach has its origins in 'The Bruntland Report' (World Commission on Environment and Development, 1987), which recognised that, particularly in developing countries, economic development and environmental protection are both required and must proceed together. The concept is more notable for having widespread support and a multiplicity of definitions (see e.g. Pearce et al, 1989) than any agreed set of associated policies. However, it is clear that any sensible definition must include not continually reducing the stock of key global resources such as forests and soil or impairing the operation of essential atmospheric and climatic systems.

Sustainable development is therefore an increasingly important paradigm for large scale changes such as deforestation, desertification, stratospheric ozone destruction and global warming. These have macroeconomic consequences and their control will require a recognition that environmental resources and services are fundamental to economic activity rather than a marginal perturbation of it.

In principle the monitoring regimes required to implement sustainable environmental regulation are not technically problematic. The relevant issues such as emissions of carbon dioxide and CFCs, loss of soil and forest resources, are measurable to the accuracy needed. The technical and scientific problems are in the understanding and modelling of highly complex systems rather than the measurements. But the greatest problems are in the political process, due to the inequitable distribution of costs and benefits of control between different interest groups, countries and generations.

Regulatory systems designed with respect to sustainable development are in their infancy. One firmly established example is the Montreal Protocol for the protection of the ozone layer. This seeks to reduce releases of halogenated compounds to the stratosphere to levels consistent with preventing further ozone loss and eventual regeneration.

Binding commitments on carbon dioxide emissions would be the key development along these lines for the energy sector. However, the UNCED meeting in Rio de Janeiro (The Earth Summit) showed the resistance of the major energy exporters and industries to the adoption of the stringent targets required (Grubb et al, 1993).

Measuring benefits - How do we do it?

For the reasons described above, the application of cost-benefit analysis to environmental regulation is becoming more important in most contexts. This poses problems for measurement, monitoring, modelling and analysis.

The calculation of values of environmental impacts requires major inputs from both environmental sciences and economics. The scientific process can be generalised as a series of operations, usually known as an impact pathway, which follow the chain of events from its anthropogenic cause through to an observable change in health or the environment. This is the approach taken in the ongoing EC/US project on external costs of fuel cycles (Dreicer et al, 1993; Eyre et al, 1993).

Figure 1 illustrates the process used for the specific example of the impacts of sulphur dioxide emissions on crops (Eyre et al, 1993). Similar diagrams may be drawn for the effects of a range of different pollutants on various receptors, that is human beings and the environment (both natural and human made).

Figure 1

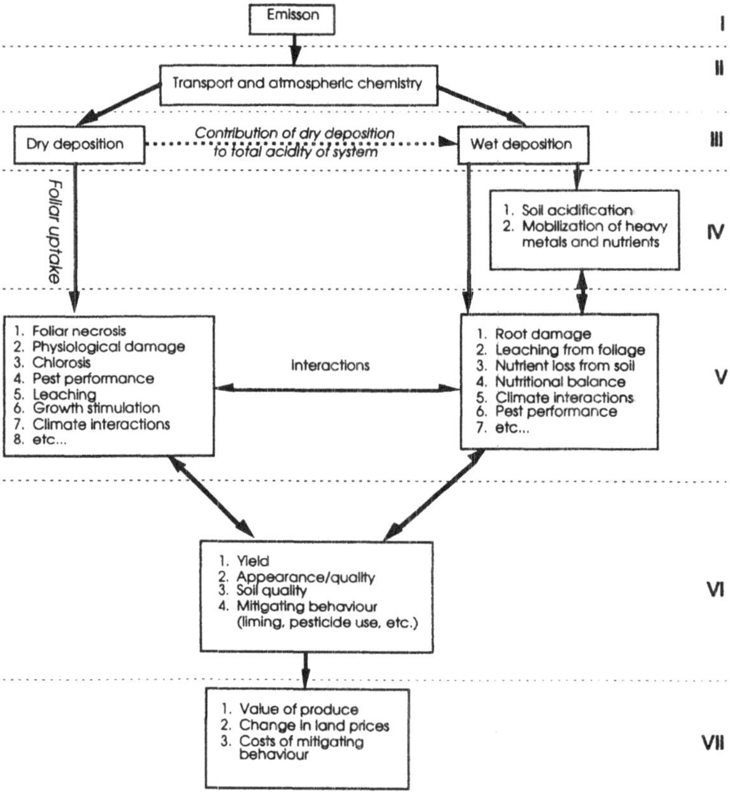

In general there is complex scientific analysis required to understand both the transportation of pollution through the environment and the dose/response relationships at the receptor. In addition a calculation of the total impact of a polluting source requires knowledge of the number and location of sensitive receptors. For long range pollution this can be problematic.

The complexity and reliability of the economic valuation depends upon the asset which is being affected. Where the damage is inflicted upon capital or goods which are traded in real markets, prices in those markets can be used to estimate value. This is relevant for damages to crops, timber, building materials etc. The theoretically correct valuation is the change in producer plus consumer surplus. However, where the losses are small in the context of overall values, it may be assumed that prices are unaffected. The product of the quantity lost and the market price is then an adequate valuation.

In many cases, however, the environmental impacts are upon human health, amenity and natural resources. These are free goods, but not without value as they are crucial factors in human welfare. Alternative methods of valuation are therefore required. A variety have been developed and used extensively, for example hedonic pricing, the travel cost method and contingent valuation. These rely on the use of linked markets, or, in the case of contingent valuation, hypothetical markets constructed in surveys, to identify monetary value.

Measuring the benefits - How good is the science ?

Extensive assessment of environmental impacts over many years has led to a vastly improved understanding of most of the major anthropogenic environmental problems. A casual observer might therefore be forgiven for assuming that quantification of the environmental impact of polluting activities should be relatively easily. Unfortunately, this is not the case. In particular, for the long range, energy related impacts (acid deposition, regional air pollution and global warming), major uncertainties remain. These come in many forms as follows:

Parameter uncertainty

In many cases the form of a relationship is known but the parameters in a dose response relationship are uncertain. For example, the rate of corrosion of some building materials under acid conditions is known to be a linear function of acidity, but the reaction rate varies from study to study, probably because of the variability of materials. This is a relatively easy form of uncertainty to handle. A range of parameter values can be derived from good studies documented in the open literature and carried through the remainder of the analysis.

Figure 2

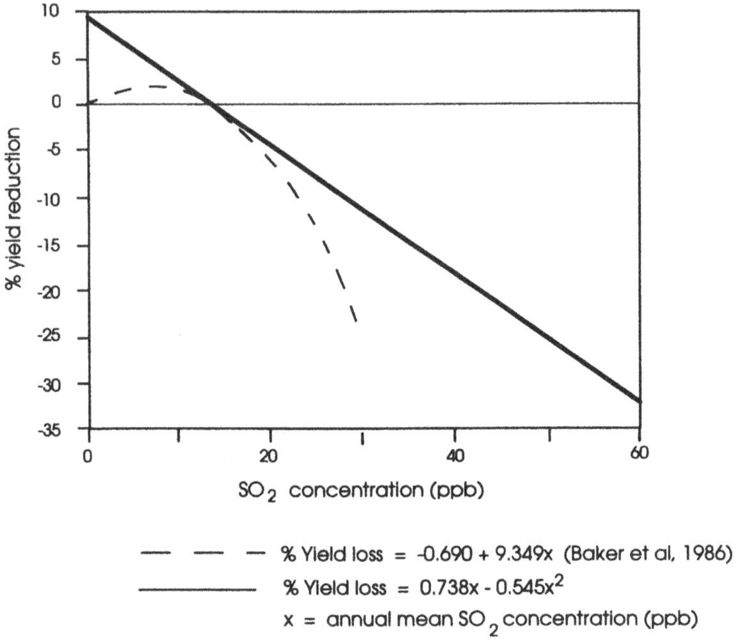

Environmental data uncertainty

Quantification of environmental impacts can require very detailed databases on the geographical distribution of receptors, people, crops, forests etc. In some cases very good geographical information systems exist. For example, land use in the UK is mapped to a very high resolution. However, in other cases data is not available in the form required and has to be estimated, introducing uncertainties. For example, for the impact of acid emissions from UK power stations on building materials, one should ideally consider the building stock of at least the whole of Northern Europe, disaggregated by location and construction materials characteristics. No database of this type exists, and therefore some fairly heroic assumptions need to be made to estimate aggregate effects.

Model uncertainty

Often the functional form of the relationship is not firmly established. Figure 2 (derived from Baker et al, 1986) provides a good example of this. Using the same data on the impacts of high levels of sulphur dioxide concentrations on barley yield, it is possible to conclude that marginal increases in pollution at levels typical of some ambient concentrations (0-10 ppb) are either beneficial (dashed curve) or detrimental (solid curve). Linearity is often assumed without strong justification. In this case there is evidence (experimental and theoretical) of beneficial effects at low pollution levels, which justifies the use of the non-linear relationship (Eyre et al, 1993).

Similar problems are common in health effects assessment. Historical data from epidemiological studies at high levels of multiple pollutants is often insufficient to identify unambiguously the effect of a single pollutant at low concentrations. In particular, there can be doubt about the existence or otherwise of a threshold below which the pollutant has no effect. Where a large pollution source disperses to give a low dose to a large number of people, this uncertainty may be critical, making the difference between a large externality and zero effect .

Model applicability

Models are frequently used outside the domain for which they were derived. The acceptability of this practice is questionable. In general, it is accepted where the underpinning theory indicates that the domain of application is sufficiently similar to the domain of derivation for the same relationships to hold. However, where the theory itself is not clear, scientific uncertainty remains. Where it is disputed, scientific judgement is required and argument can persist for a long period.

The long term effects of low doses of ionising radiation is an example. The assessments of the International Commission on Radiological Protection and others use data derived from the survivors of the nuclear weapons attacks on Hiroshima and Nagasaki. These have been extensively evaluated over the following 45 years to derive a relationship between health effects and total dose. However, the dose following the bomb blasts was over a relatively short period. The low dose received by a power plant worker over many years may give an equal accumulated dose, but with a quite different time profile. In the absence of clear medical understanding of carcinogenic mechanisms, it is difficult to be sure that the two will have equivalent effects.

In the field of ecological effects, geographical transferability is a frequent problem. The nature of ecosystems depend on climate, soil and other environmental factors, so that no two locations are identical. Some environmental impacts, such as forest decline, are observed quite widely, but many observers believe that the causes vary significantly from place to place. If this is true, simple, geographical invariant models of pollution related forest decline are invalid. Some well publicised estimates of timber related damaged (e.g. Nilsson et al, 1991) are based upon a single model derived in eastern Germany, and may therefore be unreliable.

Existence uncertainty

The most extreme form of uncertainty in environmental impacts is the question of whether the impact exists or not. Perhaps the best known example relates to the case of power line electro-magnetic radiation health effects. Although some studies show a positive effect, epidemiological evidence overall is insufficiently good to establish any firm relationship (United States Office of Technology Assessment (USOTA), 1989). However, although there is no proven medical effect, there clearly are some social effects, for example on property prices in the neighbourhood of power lines and on the location of new housing developments.

This particular example illustrates an important general point. It is impossible to design any scientific test to prove conclusively the absence of any postulated effect, the best that can be done is to establish, and then reduce, a lower limit for the magnitude of the effect. It follows that absolute safety and environmental protection are not achievable; standards can only set with reference to concepts of cost, minimum impact and risk.

Measuring the benefits - How good is the economics?

Valuing traded goods

Where environmental impacts are on goods normally traded in markets, such as crops and building materials, the valuation is a conceptually simple process based upon market prices and volumes. Where markets are distorted for some reason, prices may not represent values very well. For example, agricultural prices in the European Community are raised above world prices by the operation of the Common Agricultural Policy.

However, in general, economic modelling should allow corrections to made with reasonable accuracy.

Valuing free goods

More problematic is the valuation of free goods. For example, civilised countries aim to provide good health care free at the point of need, largely out of general taxation. However, this does not imply that good health has no value, indeed quite the opposite - it is considered too important to the community to be left to market forces. Procedures for valuing free goods, therefore, need to be considered with care.

Welfare economists have several tools designed to 'value' free goods. Where some market which is related to the free goods can be found, values can be deduced from observation of behaviour in the related markets. For example, travel costs can be used to deduce the value of recreational amenity sites, and house prices (hedonic pricing) the value of housing amenities. These techniques have been refined to a stage where they can be used with some confidence. However, there are limitations on their use. For example, hedonic pricing will measure all of the relevant amenity attributes of a neighbourhood. Determining the value of a particular pollution related benefit requires separation of other benefits affecting prices. In addition, these techniques do not measure all possible aspects of value; they only measure amenity (or user) values. Values associated with the existence of cultural and ecological assets (existence values) are not included. Yet it is existence values which are clearly the dominant concern in the protection of species, habitats and ecosystems.

Contingent valuation

Contingent valuation is the only valuation technique which can incorporate existence values, by using hypothetical markets. It is sometimes presented as a social science based alternative to natural science based 'dose-response function' approaches to valuation of environmental assets. This is a false dichotomy based on a misunderstanding of the valuation process. The thesis is that CVM studies can ask respondents questions such as 'What is your willingness to pay for the abatement of pollution from power station X?'. However, valuation of pollution without reference to its impacts is a meaningless process, because it is the impacts which are relevant to human welfare and not the pollution *per se*. To attempt to short circuit the process by avoiding the impacts stage is akin to asking someone to value a shopping basket without knowing its contents. Without any information on which to make judgements, markets, even hypothetical ones, do not function and there is no rational basis for price formation. Contingent valuation methods are therefore most appropriately applied to the output of dose-response analysis, to environmental impacts whose significance is clearly understood by the respondent.

There have been major advances in recent years in the practice of CVM to avoid some sources of bias (e.g. Garrod and Willis, 1990). Nevertheless, even where the impact to be valued is well defined, some problems remain. Ecological systems are particularly difficult. Some CVM studies have attempted to value individual species by willingness to pay to prevent extinction. But real ecosystems are complex and highly interdependent. Any attempt to value a species out of this context is therefore unrealistic. A related problem is 'embedding': CVM valuations of a sub-set of natural assets tend to produce similar values to that for the whole set (Kahneman and Knetsch, 1992). Not surprisingly, responses to CVM surveys are conditioned by disposable income, and therefore problems are encountered when it is hypothesised that large numbers of assets lose their 'free goods' status. In the limit, if all currently free goods were only accessible through markets, there would clearly be a very big effect on all prices.

Is valuation the answer? - Some paradigm issues

Valuation of the impacts of pollution tends to rely on the identification of assets which will be damaged and their individual values. This seems acceptable for most of the traditional pollution problems which are local or regional in scale. However, the 'newer' global pollution issues such as climate change throw up new problems. Expected changes include large scale desertification, flooding of whole regions and countries, mass migration and risk of war - damage categories well outside the scope of usual micro-economic analysis. Clearly for these issues it is necessary to understand interactions within the whole socio-economic systems. Ideally world environmental-socio-economic models are required in which all of the interactions can be analysed. Needless to say such models are difficult to build.

The potentially important impacts of nuclear energy are similarly problematic. Highly active waste products are expected to have their main effect on the environment only after a million years have elapsed. Large scale accidents and the consequences of plutonium proliferation are not amenable to risk assessment by statistical analysis of the frequency of past events. In other words, the problems are outside the usual domain of microeconomic cost benefit analysis.

Many impacts have very long characteristic time scales, and therefore their value as a function of time needs to be considered. The practice of discounting has been designed to ensure efficient allocation of scarce capital in real markets. Discounting of the value of free goods and discounting across generations are controversial. There is no reason to assume that questions of inter-generational equity are handled effectively within the same conventional market framework. Indeed, discounting the welfare of future generations seems inconsistent with the underlying utilitarian principle of welfare economics (Broome, 1992).

The problems in valuing key natural assets, such as major ecosystems and the services of the atmosphere and oceans are more than technical questions for environmental economists. They throw in to sharp relief the whole paradigm of placing monetary values on the natural environment. It is clearly inappropriate to allow constraints upon disposable income to determine the value of natural assets which are critical to human survival. These natural assets cannot be treated as an extension of existing market economies - they are the *sine qua non* for the operation of any economy.

Value in neo-classical economics is individualistic. Yet many feel that pollution is a symptom of a society where collective measures of value are inadequately considered - a concern neatly summed up as follows:

> 'Economists argue that all the world lacks is a suitable system of effluent taxes. They forget that if people pollute with impunity this must be a sign of lack of community.' (Boulding, 1972).

The concept of sustainable development seems to offer a way forward, in which these key natural assets are protected from significant degradation for the collective good. But the questions of which assets should be treated in this way and the level of change which is acceptable then need to be resolved.

Ultimately many of these questions are philosophical. Is it possible to reduce all human experience to a single quantifiable dimension called value? Is it acceptable to make decisions solely on the bases of human welfare without regard to other species except in so far as they impact on us? Should preferences be measured through the market or the ballot box? How should society make decisions about those impacts of environmental change which cannot be calculated or even foreseen? These questions do not have simple agreed answers. They raise moral and philosophical issues as well as purely technical ones. They certainly cannot be the province of a single academic discipline.

Conclusions

Environmental regulation is undertaken with a range of different objectives. Requiring polluters to take reasonable steps to minimise emissions, securing good environmental quality standards, maximising economic efficiency and promoting sustainable development are all possible objectives. The choice has significant consequences for the type and complexity of regulatory instrument and monitoring regime.

Measuring the benefits of environmental regulation in monetary terms is an increasingly common goal. This is difficult to achieve. Uncertainties in both the science and the economics are significant and in many cases are likely to persist for some time. Benefit estimation is therefore unlikely to provide a justifiable approach to the setting of precise levels for financial instruments such as emission fees. However, benefit estimates will increasingly provide good order of magnitude estimates, useful for identifying major environmental problems and providing guidance on BATNEEC.

For global environmental problems, the whole paradigm of valuing the environment runs up against major problems. Macroeconomic changes, intergenerational equity and environmental assets critical to human existence cannot easily be factored into conventional benefit estimation procedures. As an alternative paradigm, sustainable development provides a useful goal, but has yet to be converted into concrete policy and regulatory objectives. The problem of how to account for the natural environment in decision making is therefore not easily solved. However, this is no excuse for inaction. The scale of environmental problems related to energy use is sufficiently large to merit policy responses without every detail of a blueprint for action.

Acknowledgements

Work on 'The External Costs of Fuel Cycles', under which may of the issues discussed have been studied, is funded by the Commission of the European Communities and the UK Department of Industry. The author is grateful to co-workers in Europe and the USA for many useful discussions, and to former colleagues in ETSU for their assistance, in particular, Mike Holland for advice on biological impacts of pollution and for drawing Figures 1 and 2. The useful comments on this paper by Mike Parker are also acknowledged. The views expressed remain those of the author alone.

References

Baker, C.K., Colls, J.J., Fullwood, A.E. and Seaton, G.G.R. (1986) Depression of growth and yield in winter barley exposed to sulphur dioxide in the field. *New Phytologist* 83 627-643.

Boulding K. (1972) *New Goals for Society, in 'Energy, Economic Growth and the Environment'* (Ed S. Schurr). John Hopkins University Press.

Broome, J. (1992) *Counting the Cost of Global Warming*. The White Horse Press.

CEC (1988) Council Directive on the Limitation of Emissions of Certain Pollutants into the Air from Large Combustion Plants. *Official Journal* L 336, 7th December 1988.

Dreicer, M., Tort, V. and Manen, P. (1993) *Nuclear Fuel Cycle - Estimation of Physical Impacts and Monetary Valuation for Priority Pathways.* Commission of the European Communities. To be published.

Eyre, N.J., Holland, M.R., Krewitt, W., Mayerhofer, P., Hurley, F., Hornung, M., Markandya, A., Rhodes, B., Johnson, C.J., Prentice, A., Friedrich, R., Staiger, B., Trukenmuller, A., MacLaren, W., Waclawski, E., Robertson, A., Tran, L., Howard, P., Howard, D., Westrich, B. and Wüst, W. (1993) *Assessment of the External Costs of the Coal Fuel Cycle,* Commission of the European Communities. To be published.

Garrod, G. and Willis, K. (1990) Contingent valuation techniques: a review of their unbiasedness, efficiency and consistency. *Working Paper 10, Countryside Change Working Paper Series,* Countryside Change Unit, University of Newcastle upon Tyne.

Grubb, M., Koch, M., Munson, A., Sullivan, F. and Thomson, K. (1993) *The Earth Summit Agreements - A Guide and Assessment.* Earthscan.

Johnson, S.P. and Corcelle, G. (1989) *The Environmental Policy of the European Communities.* Graham and Trotman.

Kahneman, D. and Knetsch, J. (1992) Valuing Public Goods: The Purchase of Moral Satisfaction. *Journal of Environmental Economics and Management* 22 57-70.

Nilsson, S., Sallnas, O. and Duinker, P. (1991) Forest Potentials and Policy Implications: A Summary of a Study of Eastern and Western European Forests. *International Institute for Applied Systems Analysis Executive Report 17*, Laxenburg.

Pearce, D.W., Markandya, A. and Barbier, E.B. (1989) *Blueprint for a Green Economy*. Earthscan.

USOTA (1989) United States Office of Technology Assessment. *Effects of Power Frequency Electric and Magnetic Fields*, OTA-BP-E-53, Washington DC.

World Commission on Environment and Development. (1987) *Our Common Future*. Oxford University Press.

CHAPTER 4

A cost benefit analysis of slowing climate change

David Maddison

Introduction

There are many conflicting views regarding the true dimensions of the climate change problem. Often these viewpoints are based on different estimates of key parameters. With little indication as to why parameter estimates vary so much and which ones might be the more reliable the prospects for a co-ordinated strategy to tackle climate change appear dim. Little consensus has emerged, for example, regarding the costs of abating carbon emissions. Some take the view that considerable improvements in energy efficiency are available at negative cost by setting minimum efficiency standards for vehicles and domestic appliances (e.g. Mills et al., 1991). Others regard the macroeconomic costs of even small reductions in carbon emissions as substantial (Manne and Richels, 1993). In the judgement of other researchers, a proportion of carbon emissions arise purely as a result of institutional failings; either at a regional level in the subsidies paid to fossil fuel energy producers in the erroneous belief that this promotes economic growth (Shah and Larsen, 1994), in failure to deal with air pollution (Glomsrod et al., 1992) and vehicle congestion in urban areas (Hughes, 1994), or at a global level in failing to make appropriate transfers to those countries whose forests contain rich biodiversity which we would wish to preserve (Pearce, 1991).

But differences in opinion over the costs of abatement are minor in comparison to the disagreements relating to the potential damage if significant cuts in emissions are not made. A recent survey of expert scientific opinion (Nordhaus, 1994) indicates enormous uncertainty regarding the impact of climate change. Some see a few degrees temperature increase spread over a century as something which would hardly be noticed perhaps even benefiting colder countries in the North (Schelling, 1992). Others (see Bolin et al., 1986) emphasise the allegedly beneficial impact of elevated CO_2 concentrations on plant growth. Many ecologists, however, point to the painfully slow migration rates for some ecosystems and argue that mass species extinction would be inevitable if rapid global warming were to occur. The frequency of tropical storms could increase in line with sea surface temperatures resulting in devastation to areas prone to those kind of events (Emanuel, 1987). Some studies point to flooding of low lying island states in the Pacific Ocean and uncomfortably high temperatures in the mid latitudes.

Even the impact of escalating emissions of greenhouse gas is a matter of some debate. The impact of the elevated concentration of carbon in the atmosphere can only be resolved by reference to the temperature record. But in the opinion of some, the temperature record is itself in doubt due to thermal pollution from cities. And in any case, the observed temperature record is consistent both with the hypothesis of no climate change as well as with a 2.5°C rise (Intergovernmental Panel on Climate Change (IPCC), 1990). Nonetheless, researchers using different methods claim to have detected a statistically significant role for CO_2 concentrations (Tol and de Vos, 1994). Recently Charlson (1992) has argued that sulphate aerosols have the ability to mask the onset of climate change and that a warming rate higher than that currently predicted is likely. Some have rejected the IPCC's views altogether. For example; a poll of American meteorologists showed that none believed that any global warming had occurred so far (Lindzen, 1992). It has also been difficult to decide the extent to which changes in sea levels reflect the thermal expansion of sea water rather than vertical movements in the surface of the land following the last ice age (see Bolin et al. 1986 for a survey).

Confronted with such uncertainty regarding the costs and benefits of slowing anthropogenic climate change and even uncertainty regarding the processes underlying any observed changes, it is hardly surprising that different authorities have drawn opposing policy conclusions. Some see the costs of emissions abatement as being insignificant or even negative and advocate a rapid transition to a low carbon economy (Greenpeace, 1993). Others regard proposals for immediate reductions in carbon emissions as precipitate and strongly oppose drawing resources away from more pressing concerns just to prevent a 'slight' change in climate to which mankind could probably adapt (Beckerman, 1994).

Most economists have cast the problem of global environmental change in terms of an exercise in cost benefit analysis. A number of such analyses have appeared in the literature, for example the Carbon Emissions Trajectory Assessment (CETA) model of Peck and Tiesberg (1992). In a separate cost benefit analysis of arresting climate change Cline (1992) considers the desirability of a 4GtC emissions ceiling compared with business as usual. Both emissions reductions and sink enhancement are considered as a means of achieving this goal. Nordhaus (1992) has developed an elegant optimal control model of GHG abatement entitled DICE. This model is based around a Ramsey type model of economic growth in which all climate damage has a market impact cutting income and reducing emissions. For many environmentalists however, the considerable uncertainties regarding the response of the planet are reason enough to reject any attempt to balance the costs and benefits of slowing climate change. Instead they propose a 'carbon budgets' approach in which a historical analogue is sought for threatened rapid changes in climate (see for example Krause et al., 1989). Annual carbon budgets are set to ensure that a maximum rate of change is not exceeded.

This paper attempts to condense a diverse body of information relating to economic growth assumptions, carbon emissions forecasts, abatement cost estimates and global warming damage functions and incorporate them into a model which weighs up the costs and benefits of halting climate change. The paper does not seek to resolve the large uncertainties associated with virtually every aspect of the climate change problem. Rather, the paper asks - what do conventional assumptions about the size of the problem imply and how do changes in those assumptions affect the optimal policy? Questions regarding how one should proceed under uncertainty and what is the value of information can also be addressed within this relatively simple framework but are deferred to a later paper.

The model takes baseline economic output and future GHG emissions as given. According to the statistically estimated model of the carbon cycle used, only two thirds of these emissions are ever removed so atmospheric concentrations of carbon dioxide emissions are set to increase. Average global temperatures respond to elevated CO_2 concentrations via a process of lagged adjustment. This rise in temperature is taken as an index of global environmental change which is assumed to cause economic damage. Such losses are quantified in terms of a damage function. The warming process can be delayed by reducing emissions from the baseline (which incurs a cost) or alternatively emissions can be offset by terrestrial sink enhancement, i.e. planting forests to absorb carbon from the atmosphere. But afforestation has a lagged impact on the atmospheric concentration of CO_2 which ceases once the forests have grown to maturity. Reforestation programmes are also constrained by the costs of permanently removing land from other productive uses. The interesting question is the extent to which society ought to be willing to incur control costs now in order to prevent future climate change using rules commonly applied to project appraisal.

Clearly the model is in many ways highly stylised, neglecting potentially important feedbacks and interdependencies. This is inevitable for the construction of a fully integrated model is not feasible and in the view of the author, the following kind of 'synthetic' modelling exercise will continue to play an important role in policy analysis.

The remainder of the paper is organised as follows. Section two describes the model in more depth giving details of the data sources for the estimated equations and the properties of the equations whose role it is to describe the response of the climate to anthropogenic carbon emissions. The third section describes the results of the model when it is run using conventional assumptions regarding future economic growth rates and carbon emissions. The results contrast the present value costs of permitting unchecked carbon emissions with those associated with the optimal policy. The paper calculates the tax rates necessary to secure the optimal level of abatement activity. A variety of other GHG protocols are evaluated including measures aimed at institutional reform to halt deforestation and end the payment of subsidies to fossil fuel producers. Finally the paper concludes with a discussion of the role of uncertainty and a consideration of the scope for further simulations of the model.

The model

The model takes the form of a dynamic non-linear programme whose objective is to minimise the present value costs associated with climate change policy. In essence, the model evaluates the trade off between abatement activity and the costs of climate change related damage taking emissions reductions and sink enhancement as the control (choice) variables. This model is driven by exogenous input assumptions for economic growth and carbon emissions. The economic growth assumptions in Table 1 follow the Stanford Energy Modelling Forum specifications which suggest a slowly declining rate of economic growth over the next century (see Gaskins and Weyant 1993 for details). The emissions assumptions in Table 2 are taken from the IPCC (1992) 'best guess' scenario IS92a. Only the emissions of carbon are considered. Carbon dioxide, being the main GHG, is taken as a proxy for all GHGs. These trends in economic growth and emissions are projected forward to the year 2250 which is the terminal date in this analysis.

Table 1
Baseline GNP assumptions
($tr)

Year	*Gross National Product*
1990	22.92
2000	30.97
2010	39.77
2020	51.06
2030	66.55
2040	81.94
2050	100.89
2060	122.73
2070	149.31
2080	180.64
2090	217.35
2100	261.52

Source: Stanford Energy Modelling Forum.

The implications for atmospheric concentrations of CO_2 of these carbon emissions depend upon the dynamics of carbon in the atmosphere and the carbon cycle. The econometrically estimated model of the carbon cycle used here posits the existence of a long run equilibrium relationship between the end of period change in the atmospheric concentrations of CO_2 and the flow of carbon emissions into the atmosphere. A polynomial distributed lag was used to represent the manner in which emissions are removed from the atmosphere. The equation was deliberately overfitted and tested down.

Table 2
Business as usual annual carbon emissions (GtC)

Year	*Energy and Other*	*Deforestation*
1990	6.2	1.3
2000	7.2	1.3
2010	8.8	1.2
2020	10.3	1.1
2030	11.6	1.0
2040	12.7	0.9
2050	13.7	0.8
2060	15.0	0.6
2070	16.4	0.4
2080	17.7	0.3
2090	19.1	0.1
2100	20.4	-0.1

Source: IPCC (1992) scenario IS92a.

The data used for this exercise is drawn from a variety of sources. Atmospheric CO_2 concentrations are constructed from the Siple ice core record (Friedli et al., 1986) for the period 1744-1958 and the direct measurements taken at Mauna Loa for the period 1959 to present day. The record of carbon emissions is taken from Keeling (1991) for the period 1860-1949 and from Marland and Boden (1991) for the period 1950-1990. These records of emissions are of course deficient in that they exclude carbon emissions from land use changes which are largely unknown. In order to deal with this problem the equation was estimated with an intercept to represent a constant rate of increase in atmospheric concentrations due to land use changes. The intercept was estimated to have the value 1.147 and indicates that each decade the atmospheric concentration of CO_2 increased by a little over 1ppm independently of fossil fuel emissions. Given that the equation also suggests that ultimately 36% of emissions are retained in the atmosphere, the resulting calculation is that the average annual carbon emissions from deforestation and other non energy emissions of carbon were 0.675GtC per annum. Although a crude calculation, this figure agrees surprisingly well with the deconvolution of ice core data by Siegenthaler and Oeschger (1987) who estimate biospheric emissions at 0-0.9GtC per annum. In the equation presented below the intercept is dropped because emissions from land use changes are explicitly included:

$$CONC_t = CONC_{t-10} + 0.00446EM_{t-10} - 0.00158EM_{t-20} - 0.00118EM_{t-30} \quad (1)$$

where $CONC_t$ is the concentration of CO_2 measured in ppm and EM_t is the annual flow of carbon emissions measured in GtC. The equation suggests that by the end of ten years approximately 94% of emissions are still airborne. At the end of 20 years this has fallen to 61% and after 30 years to 36% after which time the process becomes so slow that it may for all practical purposes be taken as complete. This equation has a number of properties in common with more complex representations of the carbon cycle e.g. the model of Maier-Reimer and Hasselmann (1987). The concentration of a pulse excess of carbon into the atmosphere does not go to zero even in the long term and there is a period of extremely rapid elimination of carbon to begin with in both models. In the model of Maier-Reimer and Hasselmann however, only 15% of the carbon emissions are ultimately retained, but this process requires a period of several hundred years to complete.

Actual warming is calculated by fitting a polynomial lag with zero end point restrictions to the global temperature record constructed by Jones et al. (1991) covering the period 1854-1990. This temperature record is not corrected for the ENSO phenomenon and all factors having a transitory impact on the climate such as solar cycles and volcanic eruptions, are ignored. Once again the order of the polynomial and the length of the lag are exaggerated before testing down. The final equation suggests that the response of the climate to heightened radiative forcing is complete within forty years; Tol and de Vos (1994) reach an identical conclusion. The full impact of a doubling of CO_2 concentration leads to an estimated temperature rise of 3.5°C. Because, however CO_2 is being used as a proxy for all greenhouse gases this figure should be reduced by around one third before comparing it to the IPCC estimate for equivalent CO_2 doubling. Once this has been done the forecast equilibrium response of 2.3°C is very close to the IPCC's prediction of 2.5°C. The fact that the temperatures in the data series are measured as deviations from the 1950-1979 average means that in order to calculate the temperature rise from pre industrial levels it is necessary to adjust the intercept slightly such that equilibrium warming for pre industrial carbon concentrations is zero. The suggested temperature rise ($WARM_t$) for the year 1990 is a shade under 0.8°C.

$$WARM_t = -28.00 + 1.49CONC_{t-10} + 1.99CONC_{t-20} + 1.49CONC_{t-30} \quad (2)$$

The damage function takes this temperature rise as an index of global environmental change, calculates the amount of economic damage and scales it to GNP ($DAMAGE_t$). Only a handful of research papers have attempted to place economic values on the impact of climate change. None of these papers indicate the extent to which damage depends upon the rate of warming but instead focus upon an equilibrium climate change. It must be admitted that our knowledge in this area is dreadfully weak.

The results of a survey of expert opinion conducted by Nordhaus (1994) suggest a loss of 3.6% of GNP for a 3°C temperature rise whilst a taxonomy of climate change impacts prepared by Cline (1992) points to much smaller losses of 1.1% for a 2.5°C rise and damage increasing by a power of 1.3 with temperature rise. Fankhauser (1993) arrives at an estimate of 1.5% for a 2.5°C rise relative to pre industrial levels. Titus (1992) uses an estimate of 2.5% of GNP for a 4°C rise in temperature. The model follows Fankhauser (1993) in assuming a loss of 1.5% for a 2.5°C temperature rise and takes a damage function exponent of 2, so that the following equation is calibrated:

$$DAMAGE_t = 0.0024WARM_t^2 \quad (3)$$

In order to avoid global temperature rises, either it is necessary to reduce emissions of carbon, or to re absorb the carbon after it has been emitted. The costs of emissions reductions have been the subject of a considerable research effort over the last few years although the questions are not yet resolved because new issues continue to present themselves and there is a debate regarding the appropriateness of different modelling paradigms. There are also interesting questions regarding the costs of cutting emissions of other greenhouse gases. These have received surprisingly little attention but are not addressed in this model.

Gaskins and Weyant (1993) report that the carbon tax required to reduce US emissions to 80% of their 1990 levels by the year 2020 ranges from between $50 to $300. The costs of achieving the same reduction in emissions ranges from between from 0.9% of GDP to 1.7% of GDP. This implies a reduction in the average annual growth rate of 0.05%. These losses can be significantly reduced by using the revenues to reduce existing distortionary taxes. Dean and Hoeller (1992) conduct a similar exercise for the OECD using six different models run under common input assumptions. They consider the marginal costs of abatement curves for the years 2000, 2020, 2050 and 2100. As expected, the marginal cost of abatement curves are highly non linear. Important differences emerge when later years are considered. The explanation offered is that the existence of a backstop technology in some models implies that once the backstop technology becomes economically viable, further reductions in CO_2 are available at constant marginal cost.

Interesting though these comparisons may be, they give no indication as to which model provides the most credible results. For even if standardised input assumptions could eliminate all the disparities between the different models (and the results suggest that they cannot) this does not vindicate the modelling approach. All models could be wrong together. Equally it is obvious that no selection can be made on purely theoretical grounds - all the models are abstractions from reality and hence 'wrong' to a greater or lesser extent. Given that it will be a long time before we possess models which can be tested against the data (rather than against each other) and that the need to make policy is immediate, how should we make the best of the results that we have? One approach is to combine the results from several different models using regression analysis and to regard the estimated equation as being preferable to those from any one model. Unfortunately, there is no reason to believe that the errors made by the different models have zero mean. Nevertheless, whilst there is no justification for attaching any special significance to the results of this exercise the approach continues to find favour (Nordhaus 1991 and Cline 1992) for two reasons: it is effective in placing less weight on 'extreme' findings and 'conveniently summarises' the information contained within these models in just a single reduced form expression.

Using this technique, Cline (1992) for example, seeks to estimate a linear relationship between a percentage reduction in emissions away from the baseline and a cost in terms of a percentage of GNP using a survey of published abatement cost estimates. Notwithstanding a number of problems[1] his results seem to indicate that reducing carbon emissions by a further 10% decreases GNP by 0.9% and that delaying the cutback by 10 years reduces GNP costs by 0.25%. Cline, at a later stage, reduces these estimates even further as he accounts for the recycling of the tax revenue and associated reductions from reduced air pollution. Clearly the manner in which Cline condenses the available cost estimates into his cost benefit analysis raises questions particularly about the derivation of extent of zero cost emissions reductions, the assumed linearity of the abatement cost curve and inclusion of cost savings from revenue recycling. On the other hand, it is easy to sympathise with Cline because the published work is in many respects deficient. In these situations, judgement is required.

Table 3
Global CO_2 abatement cost modelling studies

Model	*% Cut in CO_2*	*% GNP Cost*	*Year*
Anderson and Bird	68	2.8	2050
Barns et al.	26	0.74	2005
	45	1.62	2020
	59	2.43	2035
	78	4.07	2050
	83	4.67	2080
	87	5.05	2095
Burniuax et al.	18	0.2	2000
	26	0.4	2005
	33	0.6	2010
	46	1.0	2020
	55	1.2	2030
	69	1.9	2050
Goldemberg et al.	50	0.0	2020
Manne	89	7.49	2100
Manne and Richels	75	4.0	2100
Mintzer	88	3.0	2075
Nordhaus	50	1.0	2100
Perroni and Rutherford	23	1.0	2010
Whalley and Wigle	50	4.4	2100

Source: Anderson and Bird (1990), Barns et al. (1992), Burniaux et al (1992), Goldemberg et al. (1987), Manne (1992), Manne and Richels (1993), Mintzer (1987), Nordhaus (1990), Perroni and Rutherford (1991) and Whalley and Wigle (1991).

The abatement cost estimates used to generate the regression equation used in this exercise are summarised in Table 3. All the abatement cost estimates are drawn from simulations in which it is assumed that emissions trading occurs, so costs are kept to a minimum. After some experimentation with a more general polynomial abatement cost curve a cubic equation was found to provide the best overall fit. The equation was estimated without a constant term such that the abatement cost curve passes through the origin. In other words, the equation denies the existence of any 'free lunch' reductions in carbon emissions. The scope for cost free reductions from institutional reform are considered separately. The estimated equation is:

$$ABCOST_t = 0.081CUT_t^3 \qquad (4)$$

What this equation actually entails in terms of the GNP costs of carbon emissions reductions is shown in the Table 4 and a measure of the fit of the equation to the data points can be gleaned from Figure 1. It is important to understand that the estimated equation has no particular statistical significance. Adding the results from many different models does not obviously result in a 'better' cost curve. Furthermore, the abatement cost estimates generated by the equation reflect the costs of reducing by a given fraction emissions of CO_2 from the consumption of fossil fuel, not total emissions, which include emissions from deforestation.

Table 4
The GNP costs of carbon emissions reductions

Percentage Reduction	*Percentage GNP Cost*
0	0
10	0.01
20	0.07
30	0.22
40	0.52
50	1.01
60	1.75
70	2.78
80	4.15
90	5.90
100	8.10

Source: See text.

Turning now to the re-absorption method, there has been considerable discussion regarding the potential for reforestation to sequester carbon dioxide from the atmosphere as an alternative to reducing emissions (see Cooper 1983, Brown et al. 1986, Marland 1988 and Sedjo and Solomon 1989). The key aspect of this afforestation is that only a growing forest absorbs carbon. Once forests reach maturity there is no further contribution to carbon fixation. Carbon fixation is typically modelled as being a cumulative process which is completed within 40 years.

The fundamental parameter in forestry models is the quantity of carbon contained within a hectare of mature forest minus of course, the carbon content of the vegetation currently occupying the land which is to be reforested. Marland (1988) estimates that 500mha of rapidly growing forests would sequester 5GtC annually. Woodwell (1987) calculates that 300mha remove 1.5GtC annually. Sedjo and Solomon (1991) claim that 465mha sequesters 2.9GtC annually. Nordhaus (1991) assumes that the average annual rate of removal of carbon from the atmosphere is 1.6tC per hectare in the tropics and 0.82tC in temperate areas. For the United States Moulton and Richards (1990) point to annual sequestration rates of 4.5 - 7.1tC/ha. This study assumes net storage of 186tC/ha can be achieved in both tropical and temperate regions and that the impact of afforestation is exhausted after 40 years have elapsed (since by then the trees are fully grown). The annual sequestration rate over that period is 4.65tC/ha - quite close to the average Figure found in the literature (see Table 5). A figure of 186tC/ha points to somewhat higher storage of carbon than could be achieved by converting fallow land into natural forests even in temperate regions (see Brown 1992) but this reflects the fact that the forests are managed specifically with a view to sequestering carbon. Obviously there are questions to be answered regarding the productivity of the land and how intensively the land is managed before much more can be said about the potential for reforestation. Meantime the average annual quantity of carbon sequestered by forests ($SEQUEST_t$) is estimated to be 0.00465GtC/mha during the 40 years following planting ($PLANT_t$ is the area measured in mha planted at time t) and zero thereafter. Hence:

$$SEQUEST_t = 0.00465PLANT_t + 0.00465PLANT_{t-10} + 0.00465PLANT_{t-20} + 0.00465PLANT_{t-30} \qquad (5)$$

Several studies into the potential for afforestation have sought to determine the cost per ton of carbon sequestered by dividing total establishment costs for a hectare of forests by the net carbon carrying capacity. This is usually done with the intention of comparing the relative merits of carbon sequestration versus emissions reductions. Sedjo and Solomon (1991), for example cite estimates that the plantation costs of forests are $400/ha everywhere but the opportunity cost of the land required is determined by its location. Cline (1992) conservatively estimates the costs of absorbing a tonne of carbon as $5 for tropical regions and $20 for temperate regions. Even in the United States a considerable quantity of carbon could be absorbed at a cost of less than $20 per metric tonne (Moulton and Richards, 1990). Unfortunately, making a direct cost comparison between absorption through reforestation and emissions abatement is not so easy. Reforestation absorbs carbon slowly over a period of time and the present value of a tonne of carbon absorbed in forty years time is likely to differ considerably from the value of a tonne of carbon absorbed today. Studies which seek to equate the marginal costs of emissions abatement with the cost per ton of carbon sequestered through afforestation are in effect not comparing like with like. But by explicitly modelling sequestration as a lagged function of plantation, the analysis performed here avoids such miscalculation.

Table 5
Carbon annually sequestered by reforestation

Study	*Carbon Sequestered (tC/ha/year)*
Marland	10.0
Woodwell	5.0
Sedjo and Solomon	6.2
Nordhaus Tropical	1.6
Nordhaus Temperate	0.8
Moulton and Richards	4.5 - 7.1
This Study	4.65

Source: Marland (1988), Woodwell (1987), Sedjo and Solomon (1991), Nordhaus (1991), Moulton and Richards (1990) and own calculations.

Table 6 calculates the present value costs of reforestation as a function of total land reforested. Estimates of land availability, the cost of land, management cost and plantation cost are taken from Nordhaus (1991). Given net carbon storage of 186tC/ha these estimates suggest that the present value costs of removing a tonne of carbon range from $5 to $26.

These costs are annualised and expressed as a fraction of GNP. The different geographical options are ranked in terms of increasing cost and then integrated in order to obtain the total annual cost of reforestation in terms of GNP as a function of the area planted. Regression analysis is once more used to fit a smooth curve to the abatement cost function (see Figure 2) which yields the following equation:

$$REFCOST_t = 9.935 \times 10^{-7} AREA_t + 9.132 \times 10^{-9} AREA_t^2 \quad (6)$$

$REFCOST_t$ is the annual opportunity cost of reforestation expressed in terms of GNP as a function of the area covered ($AREA_t$) measured in mha. It is assumed that the plantations are permanent.

Despite all this effort it must be admitted that little is known regarding the current opportunity cost of land in Less Developed Countries and how this might change in response to large scale reforestation. There are major unanswered questions regarding the environmental costs of large scale reforestation programmes as well as the prospects for cost recovery some years later. All this makes it difficult to assess the cost effectiveness of reforestation programmes in the manner we would wish.

Table 6
The present value costs of afforestation

Area (mha)	*Land ($/ha/yr)*	*M'ment ($/ha/yr)*	*E'ment ($/ha)*	*PV ($/ha)*
100	20	10	400	1000
100	20	20	450	1250
100	50	10	400	1600
100	100	10	400	2600
100	100	20	450	2850
100	200	20	450	4850

Source: Nordhaus (1991).

This concludes the specification of the model which is subsequently solved in step intervals using the GAMS software (see Brooke et al., 1992). Further details regarding the specification of the model are available from the author upon request.

Results

This section outlines the results which emerge from the model when it is run using the input values described above. Results are reported only up to the year 2100 to avoid any problems associated with the existence of a terminal date.

Table 7
The impact of controls on the climate

Year	*No Controls Warming (°C)*	*Optimally Controlled Warming (°C)*
1970	0.5	-
1980	0.6	-
1990	0.8	-
2000	1.0	1.0
2010	1.2	1.2
2020	1.4	1.4
2030	1.6	1.6
2040	1.9	1.8
2050	2.2	2.0
2060	2.5	2.3
2070	2.7	2.5
2080	3.1	2.8
2090	3.3	3.0
2100	3.6	3.3

Source: See Text. Temperature rise is relative to pre industrial levels.

In the business as usual scenario temperature (measured against pre industrial levels) rises from 0.8°C to 3.6°C (see Table 7). The present value cost of the greenhouse problem (Table 10) itself is estimated to be $8.9 trillion. This is the amount that we should be prepared to pay for information leading to a cost free solution to the global warming problem.

In the optimal control scenario, the optimal cutback in carbon emissions appears to be 12.7% by 2000 rising to 22.9% by the end of the next century (see Table 8). These controls on carbon emissions are rather modest. Similarly, the option of sink enhancement is a cost effective one but constrained by the availability of low cost land. It appears that those forests which have been established have been done so at a present value cost of much less than $1000/ha. Although the plantation of forests appears to be erratic, the lagged impact they have on net carbon emissions leads to an amount equal to 170 million tonnes of carbon per annum sequestered over the coming decades. It should be remembered that there are no constraints upon the rate of afforestation. However, in scenarios where a much greater cutback in emissions is called for (perhaps because damages are deemed to be high or the climate more sensitive to radiative forcing) then afforestation might yet have a role to play.

Table 8
Optimal reduction in fossil fuel emissions and forest planting

Year	*% Emissions Reduction*	*Forest Planting (mha)*
2000	12.7	37.1
2010	14.2	0
2020	15.4	0
2030	16.1	0
2040	17.8	9.2
2050	18.8	0
2060	19.7	0
2070	20.5	0
2080	21.4	0
2090	22.1	0
2100	22.9	0

Source: See Text.

With so little abatement being desirable the temperature change associated with the optimal climate policy looks very similar to the 'business as usual' (BAU) temperature change (see Table 7). In fact, the optimal control and BAU paths for temperature rise look identical to begin with and only start to diverge significantly by the year 2040 (see Figure 3). At the end of the next century the difference in temperature is still only 0.3°C less than in the BAU scenario. The optimal tax rate (Table 9) necessary to push the economy along this path rises quickly through time. In the year 2000 the optimal tax on carbon is $16.84 rising to $22.13 by

2010. But if one wished to evaluate a project like afforestation which would remove carbon at points of time in the future it would be necessary to discount these values. This requirement consequently substantially reduces the attractiveness of long term carbon removal schemes.

Even when applying the optimal control very little can be done to retrieve the losses. The present value costs associated with following the optimal policy are $8.2 trillion (see Table 10). Following the best policy available to us we can save a sum equal to $700 billion. The remedy is, as they say, almost as bad as the disease.

Table 9
Carbon tax rate ($/tC)

Year	*Optimal Tax Rate ($/tC)*
2000	16.84
2010	22.13
2020	28.52
2030	36.27
2040	45.71
2050	56.93
2060	70.25
2070	86.02
2080	104.67
2090	126.71
2100	152.76

Source: See Text. Tax rates are in 1990 US dollars.

If emissions are reduced by 9.5% in each period to represent the removal of subsidies to fossil fuel energy producers and deforestation is halted by institutional reform, then the present value costs fall to $7.1 trillion (see Table 10). In other words, regulatory policy failure is responsible for additional present value costs of $1.1 trillion. By introducing further constraints into the model it is possible to examine the present value future of income associated with various other protocols which have, from time to time, been proposed. First of all it is perhaps surprising to see that delaying the optimal policy by twenty years increases costs by only $100 billion relative to starting with the optimal policy today. On the other hand, seeking to stabilise global emissions at their 1990 levels, partly by eliminating subsidies and halting deforestation, incurs costs of $11.9 trillion whilst seeking to stabilise concentrations at their current levels is almost as bad, costing $10.4 trillion. Both of these protocols are much worse than the 'do nothing' option and very high carbon tax rates amounting to several hundreds of dollars are necessary to achieve these outcomes and large additional areas of land are required for afforestation.

Table 10
The present value costs of different protocols ($tr)

Policy	*Present Value Cost ($tr)*
Business as usual	8.9
Optimal control	8.2
Optimal control, end subsidies and deforestation	7.1
Wait twenty years	7.2
Stabilise emissions	11.9
Stabilise concentrations at 450ppm	10.5

Source: See Text. Present value costs are in 1990 US dollars discounted back to 1995.

Conclusions

This paper has presented a 'synthetic' model of the optimal control of global warming. The model can demonstrate how differing assumptions regarding the costs of abatement and the damage potential from global warming translate into particular policy implications. It is important to emphasise the tentative nature of these findings and the fact that they rest upon a specific set of viewpoints which do not enjoy universal support. Nonetheless it is suggested that exercises of this kind might play a useful role in focusing the minds of those who have to take policy decisions on our behalf.

An important limitation of this analysis is that it proceeds by replacing the uncertain parameters with their expected values. The model therefore addresses the question of what would be the optimal policy to follow if all the parameters were known with perfect certainty. But since it is definitely not the case that all parameters are known with perfect certainty the results of these exercises [are not strictly policy relevant] and may yield poor advice.

Notwithstanding these and other shortcomings, the model is used to estimate the optimal cutback in emissions and the controlled rise in temperature. This paper incorporates both emissions reductions and sink enhancement as choice variables and demonstrates that an optimal policy requires the fairly energetic use of both. Following the optimal strategy reduces the overall costs associated with climate change by about $700 billion. However, this result depends on among other things the shape of the abatement cost function and the damage function. If the reduction in carbon emissions yields significant secondary benefits or if the costs of a carbon tax can be offset by reducing other distortionary taxes then a much greater degree of abatement may be desirable.

It would not be wrong to regard global climate change as partly resulting from institutional failure. In the scenario in which deforestation and subsidies to fossil fuel producers are halted the enhanced greenhouse effect to some considerable extent resolves itself reducing costs by a further $1.1 trillion. Perhaps surprisingly delaying the use of costly measures by twenty years increases overall costs by $100 billion. Other policy objectives fare less well. Stabilising emissions at 1990 levels and stabilising atmospheric concentrations at 450ppm, even if they are partly achieved by zero cost institutional reform, are both much worse than continuing with business as usual. The message seems to be that it matters little whether carbon emissions are cut or not provided that protocols to stabilise emissions or concentrations are avoided.

References

Anderson, D. and C. Bird (1990). Carbon Accumulations and Technical Progress: A Simulation Study of the Costs. *Oxford Bulletin of Economics and Statistics,* Vol. 54, No. 1, p1-29.

Barns, D. et al. (1992). *Use of the Edmonds-Reilly Model to Model Energy-Related Greenhouse Gas Emissions.* OECD Working Paper No. 113: Paris.

Beckerman, W. (1994). In *Global Warming: Apocalypse or Hot Air?* by Bate, R. and J. Morris. Institute for Economic Affairs Studies on the Environment No. 1.

Boden, T. et al., (1991). *A Compendium of Data on Global Climate Change.* Carbon Dioxide Information Analysis Center: Oak Ridge National Laboratory.

Bolin, B. et al., (1986) *The Greenhouse Effect, Climatic Change and Ecosystems* SCOPE 29 John Wiley and Sons: Chichester

Brooke, A. et al., (1992). *GAMS: A Users Guide.* San Francisco: The Scientific Press.

Brown, K. (1992). *Carbon Sequestration and Storage in Tropical Forests.* GEC Working Paper 92-24, University of East Anglia: Centre for Social and Economic Research on the Global Environment.

Burniaux, J-M. et al., (1992). *The Costs of Reducing CO_2 Emissions: A Comparison of Carbon Tax Curves With GREEN.* OECD Working Paper No. 118: Paris.

Charlson, R (1992) *Climate Forcing by Anthropogenic Aerosols* Science No.255

Cline, W. (1992). *The Economics of Global Warming.* Washington DC: Institute of International Economics.

Cooper, C. (1983). Carbon Storage in Managed Forests. *Canadian Journal of Forest Research*, vol. 13, pp155-165.

Dean, A and Hoeller, P (1992) *The Costs of Reducing CO2 Emissions: Evidence From Six Global Models* OECD Economics Working Paper No.122: Paris

Emanuel, K. (1987). The Dependence of Hurricane Intensity on Climate. *Nature*, Vol. 326, No. 2, pp483-85.

Fankhauser, S. (1993). The Economic Costs of Global Warming. In Kaya, Y. et al., *Costs, Impacts and Benefits of CO_2 Mitigation.* IIASA: Laxenburg, Austria.

Friedli et al., (1986). Ice Core Record of $^{13}C/^{12}C$ Ratio of Atmospheric CO_2 in the Past Two Centuries. *Nature*, Vol. 329, pp403-408.

Gaskins, D. and J. Weyant (1993). Model Comparisons of the Costs of Reducing CO_2 Emissions. *American Economic Review*, Vol. 83, No. 2, p318-323.

Glomsrod, S. et al., (1992). Stabilisation of Emissions of CO_2: A CompuTable General Equilibrium Assessment. *The Scandinavian Journal of Economics,* Vol. 94, No. 1, p53-69.

Goldemberg, J. et al., (1987). *Energy for a Sustainable World.* Washington: World Resources Institute.

Greenpeace, (1993). *Towards a Fossil Free Energy Future.* Greenpeace International, Amsterdam.

Hughes, P. (1991). The Role of Passenger Transport in CO_2 Reduction Strategies. *Energy Policy*, March, pp149-160.

IPCC, (1990). *Climate Change: The IPCC Scientific Assessment.* Cambridge University Press: Cambridge.

IPCC, (1992). *Climate Change: The Supplementary Report to the IPCC Scientific Assessment.* Cambridge: Cambridge University Press.

Jones, P. et al., (1991). Marine and Land Temperature Data Sets: A Comparison and a Look at Recent Trends. In Schlesinger, M. (ed), *Greenhouse Gas Induced Climatic Change: A Critical Appraisal of Simulations and Observations.* Amsterdam: Elsevier Science Publishers.

Keeling, C. (1991). Industrial Production of Carbon Dioxide from Fossil Fuels and Limestone. *Tellus*, Vol. 25, pp174-98.

Krause, F. et al., (1989). *Energy Policy in the Greenhouse, Volume 1. From Warming Fate to Warming Limit.* El Cerito, California: International Policy for Sustainable Energy Paths.

Lindzen, R. (1992) *Global Warming: The Origin and Nature of Alleged Scientific Consensus* OPEC Seminar on the Environment: Vienna.

Maier-Reimer, E. and K. Hasselman, (1987). Transport and Storage of Carbon Dioxide in the Ocean and in Organic Ocean Circulation Carbon Cycle Model. *Climate Dynamics*, No. 2.

Manne, A. (1992). *Global 2100: Alternative Scenarios for Reducing Carbon Emissions.* OECD Working Paper No. 111: Paris.

Manne, A and R. Richels (1993). *Buying Greenhouse Insurance: The Economic Costs of CO_2 Emission Limits.* MIT: Cambridge.

Marland, G. (1988). *The Prospect of Solving the CO_2 Problem through Global Reforestation.* Report DOC/NBB-0082, Washington DC: US Department of Energy, Office of Energy Research.

Marland, G. et al. (1989). *Estimates of CO_2 Emissions From Fossil Burning and Cement Manufacture Based On the United Nations Energy Statistics and the US Bureau of Mines Cement Manufacturing Data.* ORNL/CDIAC-25, NDP-030, Tennessee: Oak Ridge National Laboratory.

Mills, E. et al. (1991). Getting Started: No Regrets Strategies for Reducing Greenhouse Gas Emissions. *Energy Policy*, Vol. 19, No. 6, p526-542.

Mintzer, I. (1987). *A Matter of Degrees: The Potential for Controlling the Greenhouse Effect.* Research Report No. 5, The World Resources Institute: Washington DC.

Moulton R. and K. Richards (1990). *Costs of Sequestering Carbon Through Tree Planting and Forest Management in the United States.* Washington DC: US Department of Agriculture, Forest Service, December.

Nordhaus, W. (1990). *An Intertemporal General-Equilibrium Model of Economic Growth and Climate Change.* Paper presented to the workshop on Economic/Energy/Environmental Modelling for climate policy analysis, Washington DC, October 22-23 1990.

Nordhaus, W. (1991). The Cost of Slowing Climate Change: A Survey. *The Energy Journal,* Vol. 12, No. 1, pp37-65.

Nordhaus, W. (1992). *Rolling the DICE: An Optimal Transition Path for Controlling Greenhouse Gases.* Cowles Foundation Discussion Paper No. 1019.

Nordhaus, W. (1994). Expert Opinion on Climatic Change. *American Scientist,* January-February.

Pearce, D.W. (1991). An Economic Approach to Saving the Tropical Forests. In Helm, D. (ed) *Economic Policy Towards The Environment,* pp239-262, Oxford: Blackwell Publishers.

Pearce, D.W. and D. Maddison, (1994). The Social Costs of Transport. In D.W. Pearce (ed.) *Blueprint 3: Measuring Sustainable Development.* Chapter 10, London: Earthscan.

Peck, S. and T. Tiesberg (1992). CETA: A Model for Carbon Trajectory Assessment. *The Energy Journal,* 13(1).

Perroni, C. and Rutherford, T. (1991). *International Trade in Carbon Emission Rights and Basic Materials: General Equilibrium Calculations for 2020.* Mimeo. University of Western Ontario.

Schelling, T. (1992). Some Economics of Global Warming. *American Economic Review,* January.

Sedjo, R. and A. Solomon (1989). Climate and Forests. In Rosenberg, N. et al. (eds) *Greenhouse Warming: Abatement and Adaptation.* pp105-109, Washington: Resources For the Future.

Siegenthaler, U. and H. Oeschger (1987). Biospheric CO_2 Emissions During the Past 200 Years Reconstructed by Deconvolution of Ice Core Data. *Tellus,* Vol. 39, pp140-154.

Shah, A. and B. Larsen (1994). *Global Warming, Carbon Taxes and Developing Countries.* Mimeograph, The World Bank: Washington DC.

Titus, J. (1992). The Costs of Climate Change to the United States. In Majumdar, S. et al., (eds.) *Global Climate Change: Implications, Challenges and Mitigation Measures,* The Pennsylvania Academy of Science.

Tol, R. and de Vos (1994). Greenhouse Statistics: Another Look at Climate Research. *Journal of Applied Meteorology.*

Whalley, J. and R. Wigle (1991) The International Incidence of Carbon Taxes. in *Global Warming: Economic Policy Responses* edited by Dornbusch, R. and J. Poterba. MIT: Cambridge.

Woodwell, G. (1987) Cited in R.Sedjo and A. Solomon (1989) Climate Change and Forests in N. Rosenberg et al *Greenhouse Warming: Abatement and Adaptation* pp105-109, Washington: Resources For the Future

Note

Φ1 The estimated regression has a negative intercept term which suggests that the initial cutbacks in emissions are available at negative cost - emphatically not a property of the models contained in his survey. This occurs because the equation is constrained to being linear.

CHAPTER 5

Environmental accounting and energy regulation

Roger L Burritt

The growth of environmental accounting

Environmental accounting is a new area of study. Its development has occurred at two levels. The term initially became familiar in national income accounting when it was recognised that GNP had defects as a measure of economic progress. Environmental factors had to be taken into account and so a supplementary form of reporting was developed by the United Nations Environment Programme, for countries to follow on a voluntary basis. In the last five years environmental accounting at the level of nations has been overtaken by a growing concern for environmental reporting at the corporate level. Indeed, it has been claimed that environmental accounting for organisations has now taken a central role in the deliberations of the world-wide accounting profession[1]. Although environmental accounting has the capacity to serve as a link between the aggregate and organisational levels, comments that follow focus on environmental accounting for organisations.

Accounting and environmental crises

Academic accountants have explored the possibility that accounting information, and the accountants who implement financial disclosure and measurement standards, are connected to the environmental crises pointed by the bio-physical sciences.

As environmental issues associated with the energy sector are critical in terms of the ability to achieve corporate sustainable development, it is not surprising to find environmental accounting beginning to focus upon the relationship between accounting information and corporate behaviour. The need to encourage corporations to change their behaviour towards the environment is vital if, in the longer term, energy resource crises are to be ameliorated.

Environmental accounting can influence corporate behaviour in three broad ways. First, in principle, published financial reports required by legislation can be adjusted to make transparent the impact of environmental factors on the costs, revenues, assets and liabilities of an organisation. These may be produced independently of the annual financial statements, as part of an environmental audit, but they could be integrated into the annual financial accounts. Such reports present several problems: annual reports are essentially reactive in nature, fulfilling a stewardship function - what the organisation's resources were and how they were used - rather than anticipating the effects of environmental damage yet to come; corporations, as industry experts, play an important role in establishing the standards which bind their own reporting, thereby raising doubts about the values expressed; financial reports disclose financial 'facts', but where environmental matters such as contingent liabilities are concerned, there is considerable uncertainty as to what these 'facts' might be; also, public disclosure of accounting information about possible environmental liabilities might generate a competitive disadvantage and open a company up to possible litigation thereby making disclosure hard to enforce.

Second, environmental accounting information can be provided to management as a direct aid to decision-making. This proactive involvement requires consideration of appropriate 'environment adjusted' investment appraisal criteria - the impacts upon expected revenues, expenditures and cost of capital used for discounting; management performance criteria, and budgets, etc.

Third, environmental accountants can place great emphasis upon non-financial information, and development of 'sustainable' accounting systems based on, say, principles contained in BS7750, the British Environmental Management Standard. Involvement in non-financial environmental information has been mooted, but could be questioned on the grounds that accountants may not possess a relative advantage when assessing physical quantities. For example, when valuing oil and gas reserves, accountants rely on specialists in other fields to measure the quantities. They then convert these to money values for reporting purposes. The same argument might be used to shun involvement of accountants in environmental audit; though the accounting profession would have a natural keenness to be heavily involved. Focus on a new 'sustainable' accounting system also presents problems. It would need to emphasise preservation and conservation of the environment, rather than bias towards investors and creditors, as found in conventional financial accounting[2]. Such a move would require a fundamental revision of the purpose of accounting, hence the need to promote a separate area of study rather than merely extending conventional accounting to incorporate environmental factors.

The role of environmental accounting in changing corporate behaviour

There are a number of lessons to be learned from the experiences in environmental law with attempts to make corporations sensitive to environmental factors. A range of strategies to change behaviour has been used both singly and in combination, with varying degrees of success[3]. Strategies, forming part of the tools of environmental lawyers, include regulation, market incentives, facilitation (and negotiation), and information. Regulation attempts to change corporate behaviour by fiat; incentives attempt to induce changes in behaviour by altering perceptions about costs and benefits; facilitation provides the opportunity for companies to change their behaviour without directly incurring additional cost; and the information strategy acts indirectly through public and political pressure brought to bear on corporations needing to be proactive in assuming environmental responsibilities. These strategies span a continuum ranging from 'harder' to 'softer' ways of changing corporate behaviour.

The workshop on 'Energy and Environment Regulation' is largely concerned with the efficacy of the regulation strategy. Environmental law, however, has found that rather than contrasting the relative effectiveness of, say, regulation and market incentives, a number of strategic triggers have been found necessary to modify corporate behaviour. Often the primary means of opening debate on global environmental issues is through a 'soft' information strategy, of which environmental accounting information could form a part. An information strategy may well precede other strategies where it is necessary to get public support for potential costly measures, as with ratification of the International Convention on

Climate Change at the Earth Summit in 1992. Alternatively, it can provide post-legislation support for a regulation strategy. For example, in a study of German and Swiss firms the United Nations found it was possible for them to disclose financial environmental accounting information segmented between current and capital expenditure, thereby reducing environmental matters to a common metric for publication in their annual accounts[4].

Environmental accounting information has the potential to become an important tool in the strategy mix available for implementing environmental law.

The dysfunctional effects of environmental accounting

It seems to be generally agreed that, at present, there is no generally acceptable form of environmental accounting for energy; but the appropriate accounting systems do need to be designed. As the design process develops, a word of warning is in order about the possible dysfunctional effects of environmental accounting.

Whether used for stewardship purposes, ex ante decision making, or non-financial back up, environmental accounting information has the potential to produce unanticipated effects which could reduce the efficacy of regulatory and environmental policy. This is the case whether policy is based on reductionist models, holistic models, or a combination of the two.

The situation is typified by the interpretative school of thought in accounting[5]. This propounds that accounting information is as much a part of defining reality as it is of measuring and disclosing it. It suggests that desired and undesired 'real' effects flow from the interpretation of environmental (or other) accounting information, as well as from the bio-physical processes being measured[6]. In these circumstances, it may well not be enough to identify, measure, internalise and report environmental costs associated with different energy sources - difficult though this might be. As accountants have, to a large extent, the ability to create the financial reality others wish them to disclose, energy and environment regulation and policy might fall at the last hurdle, depending upon whether a bias towards conservation and preservation, or a bias towards commercial investment values, is inculcated as a result of political deliberations.

References

Blake J., 'A Classification System for Economic Consequences Issues in Accounting Regulation', *Accounting and Business Research*, 1992, Vol.22, Issue 88, pp.305-321; and

Brown V.H., 'Accounting standards: their economic and social consequences', *Accounting Horizons*, 1990, Vol.4, Issue 3, pp.89-97.

See also the contrasting case by Zeff S.A. and Johansson S., 'The curious accounting treatment of the Swedish government loan to Uddeholm', *Accounting Review*, 1984, Vol.59, Issue 2, pp.342-350.

Notes

1 See Gray R., Bebbington J. and Walters D., *Accounting for the Environment. The Greening of Accountancy, Part II;* 1993 p.xi; London, ACCA/Paul Chapman Publishing for expansion of this view.

2 German emphasis is upon creditors, while in the USA and UK investors' needs are to the fore. Burritt, R.L. further explores the possibility of basing accounts on environmental law principles in 'A survival strategy for environmental accounting', *Management Research Papers,* No.18/93; Oxford, Templeton College, 1993.

3 These are reviewed in Barker M.L., 'Environmental Quality Control: Regulation or Incentives?', *Environmental and Planning Law Journal,* 1984, Vol. 1, pp.222-232.

4 UNCTC (1992, p.101) 'Accounting for environmental protection measures', Chapter IV in *International Accounting and Reporting Issues;* New York, United Nations Centre on Transnational Corporations.

5 Hines R.D. parodies the interpretative approach in 'Financial accounting: in communicating reality, we construct reality', *Accounting, Organizations and Society*; 1988, Vol.13 No.3, pp.251-61.

6 For example, see: Ingram R.W., Rayburn F.R. and Ruland R.G., Representational faithfulness and economic consequences: their roles in accounting policy; the pragmatic and ethical distinction between two approaches to accounting policy', *Journal of Accounting and Public Policy*, 1989, Vol.8, pp.57-80

Discussion: Assessing regulatory impacts

Brenda Boardman

The title of this workshop is specifically about energy and the environment, with the following objective:

> *To further the debate on the relationship between the regulation of energy production and environmental regulation in the global context.*

It is the use of energy, as well as its production, that releases the greenhouse gases causing climate change. Thus, the workshop is addressing the ways in which environmental pollution can be reduced through lower energy demand.

Energy, unlike most other traded goods, is a *derived* demand: the consumer does not want to buy electricity or gas, but wants the energy service that is produced when they are converted in a piece of equipment into warmth, light or movement. With rare exceptions, the use of less energy depends upon the purchase of more energy efficient equipment: loft insulation, low-energy light bulbs or a more fuel-efficient car. This objective can be encouraged, indirectly, by increasing the price of energy if consumers then purchase more efficient capital goods. A more direct method of achieving energy efficiency improvements is through regulating for minimum standards of efficiency (whether in appliances or buildings) or capital grants. In this way, the consumer is able to save on running costs, without first suffering from increased prices. The latter route causes considerable hardship for lower-income groups, as demonstrated by the concern over the impact of value added tax (dealt with below).

A further method of reducing pollution, particularly in electricity generation, is to clean up the process or by utilizing renewable resources.

The three papers in this section of the book, represent a continuum: a case study of past experience in Japan; an analysis of the regulatory framework; and a theoretical economic model. All three papers underline the problems of using economic factors to determine environmental policy.

Mr Kuroki explains how the Japanese approach to sulphur dioxide control can be used as a model for emission reduction policies. Where serious health hazards are involved, Japanese public opinion is in favour of safety standards rather than a cost-effective approach. Once politicians have set the standards, the industries respond first by switching to less polluting fuels, then by introducing cleaner technology. The cost of these changes is seen as 'relatively small': a 90% reduction in both sulphur dioxide and nitrogen oxides resulted in a 45% increase in electricity generation costs. For individuals suffering from ill-health related to sulphur dioxide emissions, financial compensation was provided by a levy on the local polluting industries. The amount of the levy on a factory was proportional to its emissions, creating a further incentive to undertake investment, particularly as more people claimed. The concern about the size of the levy eventually stimulated more precautionary investment than the regulatory standards.

The Japanese experience with sulphur dioxide, Mr Kuroki believes, demonstrates that controlling carbon dioxide emissions is feasible and appropriate: the same uncertainties exist over the causal relationships and economic impact and the responses will vary over time as technology and policy develop. Meanwhile, a more healthy environment would be created for local and global residents.

Nick Eyre's paper provides an analysis of environmental regulation and examines the four main strategies. The new objective of sustainable development will not be achieved by the historical approach of negotiation between polluter and regulator, nor by safety standards related to individual harm. The uncertainties involved in valuing environmental effects means that assessments of economic efficiency are difficult, as demonstrated by an analysis of sulphur dioxide emissions on crops. Thus, the challenge of sustainable development has yet to be converted into concrete policies and regulatory objectives, but will require a new approach.

This paper provides a useful and thoughtful overview of the main forms of environmental regulation. He spans the range from the top-down approach of setting environmental objectives first, and then looking at methods of achieving them, to the bottom-up approach of imposing economic penalties on individual polluters to reflect externalities in the hope that this will result in sufficient reduction in average levels.

All modellers walk a tightrope between the simplification necessary to accommodate a wide range of variables and the risk of making assumptions that are farcical in their naivety. The courage to undertake the work has to be accompanied by a good dose of common sense if the results are to be acceptable. Unfortunately, Dr David Maddison does not encourage confidence with some of the limitations that he places upon himself:

- 'the economic agents ... are endowed with perfect foresight', but there is no consideration of renewables or nuclear power. The model only accommodates the three main fossil fuels: oil, coal, gas;
- the economic agents follow private self interest, so that 'no forests are planted since private individuals are unable to capture the benefit', presumably because they do not live long enough and private self interest does not extend to future generations;
- there is a 'population ceiling' of 11.31 bn people, implying universal birth control ;
- the cost of fossil fuels is not affected by depletion 'due to a lack of information';
- although soil is effectively a finite resource, 'the quantity of land for productive purposes evolves'.

As a result, he accepts that the uncertainty attached to the various parameters make the results 'not policy relevant'. Perhaps the early stages of economic modelling of complex issues have to progress on the basis of such simplifications. Certainly, the array of 38 equations gives the feel of a weighty paper. Sadly, the net effect for the non-economist is of a paper that lacks both conviction and clarity. The concept that economic assessments of this sort could determine the 'optimal' level of global warming is unconvincing. Concern for the environment and political decisions provide more comprehensible approaches and thus are more likely to be the route for determining emissions targets.

The potential severity of climate change has only been recognized over the last 10-15 years. As both Nick Eyre and David Maddison have demonstrated, there is a large and recognized level of uncertainty in our assessment of these environmental impacts. There is no uncertainty about the fact that climate change is being induced by our activities, merely uncertainty about the rate and impact of that change. Should we aim for precision in valuing and assessing externalities when there is this uncertainty and perhaps unknown other effects? Should we just take a broad approach, to stimulate action as quickly as possible? Even when we can identify the externalities, there are real uncertainties about what the internalised costs should be, particularly in different regions of the world, as Nick Eyre identified.

The evidence of the effects of both the Montreal Protocol on CFCs and the Rio Framework Climate Convention on greenhouse gases is that simple, universal targets are influential in changing policies. The evidence from Japan is that public concern over the health hazards of environmental pollutants provides sufficient basis for action, despite the scientific and economic uncertainties.

PART 3

Energy Pricing

CHAPTER 6

Energy pricing: Trade in energy and its impact on the environment

Nicholas Hartley

Introduction

This paper draws together two different developments, which coincidentally have come together:

(i) shifts in the structure of the energy industries - notably in the electricity supply industry - driven by the process of privatisation. There has been a move from centrally-directed monopolistic markets to more competitive arrangements, with a greater use of prices to determine the allocation of resources, and varying degrees of competition or contestability.

(ii) the changing perception of the value of environmental resources - in particular the recognition of the potential damage done during the extraction, processing and use of energy.

The central theme is the link between energy pricing and the environment. One such link is that between environmental regulations and energy prices - regulations which control the scale of harmful emissions tend to raise the cost of doing business, and so to increase prices. But governments and regulators have also put surcharges on energy prices to cover environmental concerns, either to provide the funds for environmentally-desirable expenditures or to 'correct' prices so that they reflect environmental costs, with revenues flowing into the general exchequer. The link back to developments in energy markets is that growing competition for sales, and increased international trade in energy, raise the possibility of improving the efficiency with which resources are used, including environmental resources.

Competition

Governments have traditionally played a major role in determining both the level and the structure of final energy prices. Intervention was constrained by certain fundamentals, such as the need to cover costs, and the physical possibilities for substituting the use of one fuel for another. But within these constraints prices often diverged from the levels and structures which would have been expected under competition. To put it at its simplest, governments have had other objectives as well as that of securing economic efficiency, and they have been willing to use energy prices to secure those objectives - examples relate to the distribution of income, and the competitiveness of energy-using industries in international markets.

The process of privatising the gas and electricity industries, and the introduction of competition into parts of the market, has raised fundamental questions about the cheapest way to supply energy, and about the best way to recover costs from different types of customer. The culture of the energy industries has been engineer-led. The shift towards a market-led, more economics-based, approach has been a profound one. To quote a Norwegian study of electricity markets: 'The general trend .... has been to put relatively less emphasis on issues like system dimensioning and reliability, security of supply, contract coverage of producers and distributors, technical standards, standardisation of entities of the system from a system's perspective .... and to give more emphasis on market structure and market behaviour, competition strategies, internal efficiency of individual agents, transparency of tariffs and other delivery conditions'.[1]

In the UK, the process of privatisation - and the introduction of a range of new commercial relationships - has encouraged the various parts of the industry to cut costs. But there have been more fundamental changes in the way in which the energy industries do business. The increasing sophistication of the contractual arrangements between buyers and sellers has led to greater emphasis on the development of prices over time. The aim must be to signal longer-term conditions of supply *via* the price mechanism.

Jim Skea has argued that a competitive industry is less likely to be locked into existing technologies.[2] He suggests that private investors will tend to shift investments towards technologies which are less capital-intensive than those which the old state monopoly would have chosen. He suggests that 'Lower capital intensity also implies a greater use of high quality fuels as opposed to investment in end-of-pipe clean-up technologies'. We might add that while new technologies need not necessarily be cleaner than those they replace, users know that they are living in a world of heightened environmental concern, and that the environmental regulators are likely to continue to exert upward pressure on environmental standards. In times when environmental regulations are expected to tighten, firms are likely to see it as more risky than before to invest in 'dirty' technologies. They may even start to anticipate future environmental constraints. For example, recent work by Manne and Richels at the US Electric Power Research Institute suggests that the prospect of a carbon tax (or a similar constraint) may already have started to alter investment patterns, so that all forecasts of carbon emissions should now be lower, even in the absence of immediate government intervention of this kind.

Final energy prices are composed of prices charged at a number of different stages - mining or drilling, shipping, generation, transmission, marketing, billing etc. Throughout this chain, firms have been encouraged to look at both the level and the structure of their prices. As an example, the scope for transmission prices which more nearly reflect marginal costs is particularly interesting. Sally Hunt and Graham Shuttleworth start a recent article on transmission pricing thus: [3]

'Transmission pricing in electricity is undergoing a conceptual revolution. It is now widely accepted, on both sides of the Atlantic, that marginal costs are the appropriate basis for pricing both regulated and competitive generation, and that they vary by time of use. The new transmission pricing involves expanding our view beyond time of use, to place of use. The advent of competitive markets is focusing attention on locational aspects of pricing, and the debate is now drawing on familiar concepts, such as opportunity costs and the marginal cost of transfers between nodes of a network system.'

While better pricing holds the prospect of reducing the scale of the resources used by each network, competition may lead to some duplication of networks - as in some parts of the US electricity system, and some parts of the UK continental shelf. But it should ensure that users are faced with prices which more nearly represent the actual costs of transmission. This implies that prices will reflect marginal costs, though there is an important debate about the respective roles of short-run and long-run marginal costs. In these circumstances, there should be good incentives to reduce transmission losses and to relieve transmission constraints. Hunt and Shuttleworth say that ideally new users should pay all the extra costs carried within the network as a result of their entry, with no averaging. Until now tariffs have reflected only the proximate costs associated with new demands ('shallow entry/exit costs') - though the National Grid Company (NGC)'s transmission tariffs do, in fact, now have some locational differentials.

With economic pricing, a new generator seeking entry to the grid, would receive less favourable terms if there was a transmission constraint associated with taking supplies from its plant. Equally, if the grid operator found it cheaper to upgrade the system, or to build new lines, than to take in new supplies it should be able to profit from this alternative. Using parallel arguments, incentives will be best distributed if the grid operator bears the cost of transmission losses, buying the energy lost in transmission from generators and charging for this in its contract prices.[4]

This is mainly an indication of possible developments to come. The relevance to the concerns of this paper is that, by producing a more efficient transmission system, better pricing can, in principle, reduce the need for new transmission lines or new generation. To put it rather baldly, as Sally Hunt says in another paper, in the past 'by violating the basic rules of the economics, we may have ended up violating the landscape unnecessarily'.[5]

But competition affects not only choices about the best inputs into energy production, but also the final price of energy. If environmental damage is directly related to the scale of energy use, and if competition reduces prices, because they fall towards costs, which might themselves be falling as a result of privatisation, then this might be to the detriment of the environment, since demand is likely to increase. The answer is, however, to tackle the environmental issues directly, e.g. by the imposition of taxes on pollutants, rather than by trying to hold back the growth of competition (see section on environmental regulation).

In some cases, the existing balance of prices may be the other way round - with prices at some times of day below marginal costs. In these circumstances, it may sometimes be difficult to increase prices. Consumers will be faced with the signal to use too much energy. It is too cheap. Given that the environmental gains from energy efficiency - notably the reduction in CO_2 - are increasingly seen as the prime motive for action, the inefficient use of energy has wider implications.

Any utility which, for some reason, is unable to charge enough for a service to cover its marginal cost, and which is therefore losing money on each marginal unit sold, should prefer to pay its customers to reduce demand. It is sometimes suggested that a utility should offer a subsidy of up to 5p if it is making a loss on sales which cost it 5p to produce. This is not so: the maximum amount it should pay to reduce demand is the amount it is losing by meeting that demand, i.e. the amount by which marginal cost exceeds price.[6]

One area where, in the UK, there may be scope for a demand-side response is that of bidding into the electricity pool. Industrial customers can already reduce their bills by rescheduling demand away from times of high pool prices ('demand-side management'). But users cannot bid demand reductions, i.e. promises to switch off, into the pool ('demand-side bidding'). If they could, the pool managers might sometimes prefer to pay them to reduce demand rather than paying the generators to increase supply. The detailed mechanics of such a scheme are currently under consideration. Caminus Energy, the authors of a recent report on demand-side measures, look briefly at the potential environmental impact. They suggest that:

'Widespread use of demand side measures would lead to a reduction in peaks in demand, and a flatter demand profile overall. This would in turn lead to greater utilisation of most existing generation plant and a reduced likelihood that low efficiency peaking plant would be called upon for generation. Higher plant load factors would tend to lead to improvements in overall efficiency, with resulting environmental benefits. However, to the extent that customers may pay lower electricity prices on average when demand side measures are introduced, electricity consumption may increase, depending on the customers' price elasticity of demand.'[7]

This gain reflects the averaging inherent in any tariff.

To summarise this bit of the argument: if there are environmental benefits from privatisation and competition in the energy industries these are most likely to arise from the introduction and improvement of new commercial relationships like those between the Regional Electricity Companies (RECs), suppliers, generators, and the NGC. The effect of privatisation, accompanied by restructuring, is to extend the use of the price mechanism into areas in which it previously had little role. To the extent that competition has lowered prices, then this will have been to the detriment of those parts of the environment where damage is directly related to the scale of energy use. If there is a related environmental problem, this needs to be addressed directly.

International trade

The process of competition, both nationally and internationally, is slowly starting to integrate previously separate national energy markets: ultimately there is the expectation that, as in other markets, competitive forces will seek out the most efficient pattern of production and distribution, both at home and abroad.

This leads us to one of the central environmental debates of the moment - the debate about 'trade and the environment'. The issue is straightforward: is the development of free trade compatible with the protection of the environment? At its root, this too boils down to a debate about prices and resource allocation. If environmental resources were correctly priced to take account of environmental externalities, then free trade, coupled with the proper functioning of capital and labour markets, should allow production to be distributed throughout the world in the best way on the principle of comparative advantage.

Few parts of the economy have such an intrusive impact on the environment as the energy sector. Even so the net impact of greater trade in energy may be favourable to the environment. Trade has a key role in establishing the opportunity cost of domestic energy supplies. For example, with the opening up of the economies of Eastern and Central Europe, these nations' fuel resources are now, for the first time, becoming valued at world prices. These prices are often way above the previous prices set within the planning system. There are, therefore, incentives to cut waste and profligacy. The environmental gain - e.g. the reduced emissions of global warming gases associated with reduced wastage in the Russian gas pipeline - is widely recognised. There is also a growing recognition of the importance of removing energy subsidies, on electricity, natural gas and coal. Such subsidies are substantial and widespread.[8] The reasons for these subsidies are various - but the costs of this misallocation of resources are likely to be substantial.

Currently, some energy is traded: some is not. Energy supplies, like oil and coal, which can be fairly easily transported are traded on world and regional markets. But there are now possibilities for new energy trades - in particular, while some gas and electricity is already traded internationally, there are pressures for more market entry than is now possible. European energy markets are entering a period of rapid change with new suppliers challenging old monopolies. The growing continental trade in gas, combined with concern about the scale of energy price discrepancies, are the driving factors. Pressures within the European Union, in particular, are leading towards greater trade of energy supplies within EU countries. The proposals for some form of 'third-party access' to EU energy markets would allow entry by suppliers from one country into the markets of another. The result should be that energy will be supplied from the cheapest source, which, since broadly equivalent environmental regulations are likely to be in force, may also be the source which imposes the least environmental damage.

But if regulations are not broadly equivalent, the net impact of extra trade in energy on the world environment cannot be known. The best reason for expecting that adverse environmental effects can be avoided lies in the progress of the environmental debate. Purchase of electricity or gas from overseas is likely to entail some extra investment in the network, notably in an interconnector between separate networks. There is also the risk that the price of imported energy will increase if environmental restrictions are expected to tighten. This suggests that buyers may be wary of buying imported energy if it is produced by means of highly polluting processes. Thus, even if current environmental regulations are inadequate, expectations that standards will rise are likely to put a limit on the scale of any switch to cheaper fuels (where this is the result of different environmental regulations).

There may be more concern about those fuels which do not involve substantial investments in dedicated facilities. The fear is that the process of competition, which is associated with freer trade, may lead to a decline in environmental standards, as prices are bid down. In practice, oil prices are currently at a level which is sufficient to allow most countries to impose the necessary environmental protection requirements on the oil companies (*vide* the recent environmental requirements associated with the 14th round of the bids for the UK continental shelf). Even so, freer trade in oil could, in principle, encourage countries which are marginal producers to meet world prices by failing to impose strict environmental controls.

These are, of course, choices which sovereign countries must make for themselves. Free trade allows countries to benefit from the sale of resources which they might otherwise be unable to use. Where natural resources are traded, there is always the risk of associated environmental damage. Countries at different stages of economic development will make different trade-offs between today's wealth/welfare and environmental protection. Lower environmental standards in developing countries are of direct concern to all if the environmental problem is global - since there is no sense in transferring the source of global pollution from one country to another. International concerns are different where the environmental problem is local or regional. In these circumstances, there are no direct reasons for wanting to limit the extent of pollution in another country. But public opinion already sometimes rejects trades which depend on low environmental standards, and importers are increasingly sensitive to this pressure. The extent to which the framework of international trade should seek to take account of all kinds of environmental damage is one of the central issues in the current trade and the environment debate.

To conclude, there are fears that trade liberalisation will be bad for the environment. In large measure this concern arises from a recognition of imperfections in domestic policies which mean that prices are too low to reflect all environmental costs. It is for individual countries to make sure that these costs include the unmarketed costs of environmental resources. But firms' general expectation of rising world wide environmental standards make them wary of short-term gains based on lower environmental performance .

Environmental regulation

We turn now to the impact of environmental policy on energy prices. The first task is to understand the way in which decisions are made about the scale of the environmental improvement to be sought. The starting point must be a good understanding of the underlying science. But the identification of an environmental impact must be followed by a judgement, either implicit or explicit, about whether the benefits of avoiding the damage are greater than the costs of stopping or controlling the emissions (or whatever). It is difficult to put a monetary valuation on environmental damage, but that is ideally what is required.

Electricity production, for example, impacts on the environment in the following ways: the construction of new generating sets takes up land and blights visual amenity; the process of generation by means of fossil fuels creates harmful emissions of the acid gases (SO_2 and NO_X) and of the global warming gas CO_2; the distribution network uses the resources of the countryside. We cannot be sure that all these costs are fully reflected in prices.

A first thought might be that they are currently wholly omitted from the costs of production and distribution, and so from prices. This is very far from the case: planning restrictions affect the cost of constructing new generating sets and transmission facilities; restrictions imposed under EU Regulations, and administered by the pollution inspectorate (HMIP), affect the amount of acid rain gases which may be emitted, and so the costs of generation. What is at issue is whether these costs have been correctly internalised in decisions. This may be examined by specifying what to the economist would be the 'first-best' system. In the case of amenity costs, a price should in principle be placed on the reductions in amenity associated with visual, and any other development, blight. Producers would then decide whether to bear that cost. In the case of emissions costs, a price would be attached to emissions damage - with the price varying as marginal damage varied. Producers would then decide how far to bear that cost, and how far to save costs by reducing emissions (eg by fitting control equipment or switching fuels).

The decisions which producers would come to might be exactly the same as those which were arrived at *via* the administrative (regulatory) route - but there is no reason why they should be. Where the regulator has made a special effort to establish the costs and benefits of each course of action - and to quantify these costs and benefits - then the final decisions might be close. But in practice, decisions are likely to have been taken on a much less quantitative basis. Moreover, there is one essential difference between the two outcomes - this is that where a charge is made, the user of the environmental resource continues to pay for residual pollution, even after the optimal level of pollution has been determined. To the economist this has the very great merit of correctly determining the size of the polluting industry - an influence which is absent if this income effect is missing as with the administered cost-benefit solution. But it may be an unpopular feature with industry.

The ideal requirement is straightforward - an addition should be made to every price to account for these environmental externalities (though sometimes the addition may be insignificant). These surcharges are sometimes known as 'adders'. The difficulties are two-fold: how to measure the externalities; and how to impose the surcharge.

I do not propose to go into the question of measurement and valuation in any detail - that is the subject of a separate study. There is a surprisingly wide range of techniques which can be used to value environmental impacts - ranging from estimates of pollution damages on buildings, crops, even human health, to values ascribed by the people to the preservation of certain species of wild-life. Some means of valuation are better than others. Professor David Pearce has been working on one set of adders for the UK (his first results were published in a report to the DTI[9], but these have been subject to revision[10]). The damages covered relate to acid rain damage, to global warming damage and to nuclear risks. All such estimates are subject to a range of uncertainties. An article by Andrew Stirling shows the wide range of variation in reported environmental externality estimates for coal-fired electricity in the USA. [11]

This work is most useful in showing how far the choice between different fuels would be altered by the application of adders. Stirling shows that in most cases the ranges of reported environmental externality estimates for the various electricity supply technologies overlap. Given these uncertainties, Friedrich and Voss conclude that 'the resulting external costs do not significantly change the present competitive position of electricity production from coal, nuclear power, photovoltaics and wind'.[12] Others believe that the results produce clearer answers. But there is still much work to do.

Supposing we do want to apply adders - how would we do so? This depends in part on the way in which energy is supplied. In a centrally planned system the shift could in principle be made simply by changing the planners' internal prices. Indeed prior to the collapse of the Soviet and Eastern European planning systems there were facilities within the planning mechanisms for imposing just such surcharges (though they were rarely used).

The US system of utility regulation suggests another possible route. Where energy is produced by a monopoly, it is open to the regulator to direct the monopoly to alter its own internal decision rules so as to take environmental consideration into account. For example, in the electricity industry it is possible to think of altering the computer programmes which determine the order in which generating plants are run ('the merit order') so as to minimise the output of pollutants. There are examples of this approach in the USA. As a result, costs will be increased, and the regulator must be prepared either to compensate the firm with increased revenues from other sources, or to accept that the overall level of profits will fall as a result of its actions. It is impossible to avoid a system which distorts commercial decisions.

A third option is to impose a regulation setting out the emissions limits which are allowed from particular plants. In principle this option meets the basic objectives. Regulation which is capricious, unexpected and arbitrary is likely to increase the risk of doing business, raising the cost of capital and so final energy prices. But if regulations are planned, particularly in relation to companies' and industries' investment cycles, then these dangers are reduced.

The costs of such an approach might also be reduced if pollution permits could be traded between emitters. Such a scheme would allow a government to be sure that it had reached its overall emissions target, but would give the private sector some flexibility about who would bear the reductions needed to meet the target and when. (Polluters who could reduce emissions cheaply would find it worth their while to sell permits to other firms who would otherwise find it costly to cut back.) Systems of tradable permits do, however, require a prior choice to have been made about the level of pollution which will be allowed.

The next option is arms length intervention by government *via* the tax system. This has received most attention in recent years - e.g. with proposals for a carbon tax or a sulphur tax. The idea is simple - the imposition of a tax per unit of production to reflect the environmental damages which are not measured in input prices (costs). The practical choices are much more complicated: consider the following difficulties:

(i) *how to set the tax* - the tax should reflect marginal damage, but if the tax alters the pattern and scale of production this will change - how can this be allowed for (e.g. will the tax be on a sliding scale).

(ii) *where to levy the tax* - should the tax be imposed on inputs (e.g. on units of coal), on outputs (e.g. on units of electricity sold to distributors), or on final sales (e.g. on electricity bills)? This is partly a matter of practicality (the need to define as few tax points as possible). But some options seem to be much more closely identified with the original environmental concern than others - in particular a tax which associated with fuel inputs generally seems better adapted to the environmental needs than one which is levied on outputs, since it is seldom possible to levy an output tax which varies with the composition of inputs or which encourages conversion efficiency.

(iii) *what to do about imports and exports* - if other countries have not imposed an environmental tax then the competitiveness of UK producers may suffer. One answer is to suggest some form of border tax adjustment (as was initially suggested as part of President Clinton's original BTU (British Thermal Units) tax). But this would be very difficult to administer. (For example, how do you determine the amount of carbon 'embodied' in the production of a South Korean Car?) It can be argued that so long as other countries are taking equivalent measures to meet the same problem, then their industries are suffering an equivalent loss of competitiveness. But it is unlikely to be the same industries which suffer -

which is an important political consideration. Moreover, even if all countries impose the same tax - the net impact on competitiveness will depend on the way in which the tax revenue is recycled, and this is bound to differ.

(vi) *single market considerations* - these arguments are somewhat different in the context of the single market since where energy is traded the two countries involved will wish to avoid double taxation, as would happen if one country had an inputs tax and the other an outputs tax. In principle, border adjustments could allow for this - but again they may not always be easy to administer.

Economists remain attracted to these externality correcting taxes. Others are more doubtful. Thus, despite a general acceptance among policy makers of the case for economic instruments, and despite some countries' willingness to introduce such charges unilaterally (e.g. Sweden and Denmark), many people continue to prefer other means of addressing common environmental problems. Why should this be? The following seem to be the major concerns: worries about international competitiveness and costs of adjustment generally; worries about the ways in which the money will be recycled; worries about the future level of any tax; dislike of the 'income effect' associated with the charge for residual pollution. Industry often seems to prefer the relative certainty of an administered solution - whatever its theoretical inefficiencies.[13]

In principle, the correct basis for a carbon tax is an associated estimate of a global warming 'adder'. Andrew Stirling takes comfort from the fact that today's US adders (mainly for acid rain) are generally small. He suggests that 'Values much higher than this would not be directly usable, since the effect of increasing electricity prices by a factor of 10 might be thought to be crippling to the economy and therefore inconceivable in policy terms'. And yet, if the long term carbon problem is confirmed as a result of the further work of the scientists advising the UN Inter-Governmental Panel on Climate Change, it is hard to see how the price mechanism could fail to be in the forefront of any solution. The academic literature suggests that carbon taxes of well over $100 a tonne of carbon would be needed in the next century.[14] Some existing taxes, e.g. those on petrol, when expressed in terms of $/tonne of carbon, are already larger than this (though since such taxes have other purposes this does not imply that they should be reduced).[15] The truth is that we are all used to current structures of relative prices, including the tax component, and any change would be painful.

A final option is to offer a subsidy to the less environmentally damaging alternatives to conventional generation (as with the renewables component of the Non-Fossil Fuel Obligation) or to investments in energy saving (as with the activities of the Energy Saving Trust). The subsidy should ideally represent the difference between the marginal private cost of provision and the marginal social cost of provision - so that, for example, while the marginal private costs of generation by wind power may exceed the marginal private costs of conventional generation, the subsidy represents the different marginal social costs (with wind power having fewer external disbenefits than conventional generation), so that the two are put on a more equal footing. In practice it is difficult to achieve so pure a solution.

Conclusion

This paper has ranged wide and has picked up a number of strands. These are that competition affects prices, and produces changes in the use of resources; that regulation adjusts prices so that they reflect the use of the environment as a resource, in which case the changes induced by competition can benefit the environment; and, finally, that prices affect the level and composition of international trade, which in turn has an effect on the environment. Thus, so far as environmental protection is concerned, prices matter - especially relative prices. Ultimately environmental problems are economic problems which are susceptible to classic economic remedies, like correct resource pricing.

But while there are examples of implicit environmental valuations by way of a variety of regulatory constraints and targets, there are relatively very few examples of explicit environmental valuations in practice. The issue is whether this simply reflects the fact that the 'science' of valuation is still in its infancy - or whether the task is an impossible one. Until policy-makers are satisfied that they are working with robust valuations, the basis for any decisions to override market choices will remain essentially political.

Notes

1 *Markets for Electricity: Economic Reform of the Norwegian Electricity Industry*, Einar Hope, Linda Rud and Balbir Singh, Bergen Centre for Research in Economics and Business Administration Working Paper No. 12/1993.

2 *Ownership, Market Structure and Environmental Strategy: Electricity Generation in Britain*, Jim Skea, *mimeo*, 1993.

3 *Electricity transmission pricing: The new approach*, Sally Hunt and Graham Shuttleworth, Utilities Policy, April 1993.

4 Report on Constrained-On Plant, OFFER, October 1992.

5 *The Problems of Transmission Pricing*, Sally Hunt, paper presented to *The Economist* conference The New Electricity Market: Challenges and Opportunities, 18 March 1993.

6 Let us suppose electricity sells at 3p a unit, but costs 5p a unit (at the margin). The least-cost planning literature often suggests that in these circumstances a utility should be willing to pay up to 5p to those who promise to reduce their demand by a unit. If 5p is offered to those who reduce their demands, there are both distributional and efficiency effects. The simplest way to see the distributional effect is to recognise that if users are to be paid to reduce their demand, then somebody must provide the money to pay them. In effect, this money has to be raised by a surcharge on those customers who remain with the utility. The efficiency point is this: the price of 3p represents the marginal utility of consumption. After the introduction of a subsidy, this choice is distorted: if the subsidy of 5p is given, then, in effect, the consumer will commit up to 8p of his or her own resources to provide alternative supplies (ie the 3p that would have been spent on the supplies of the utility, plus 5p from the subsidy) so that alternative means of meeting demand costing, at the margin, 8p are chosen - while the utility's resource cost of producing electricity was only 5p. If price is less than marginal cost, then a subsidy limited to this difference provides the right signal. If price is set at marginal cost, price alone provides a sufficient incentive to users to reduce their demand. To summarise: if price signals are distorted, and prices do not reflect marginal cost, there may be gains to be had from some second-best system of pricing. But the rule set out above gives the limit to such payments. In the above example, if a subsidy of 2p is provided (the difference between price and marginal cost), consumers will search for alternatives costing up to, but not beyond, 5p (the cost of conventional provision).

7 The benefits of demand side measures: a report commissioned by the Chemical Industries Association on behalf of the Energy Intensive Users Group, Caminus Energy, January 1993.

8 *Commercial energy subsidies in developing countries: opportunity for reform*, Mark Kosmo, Energy Policy, June 1989.

9 The Social Costs of Fuel Cycles, Report to the DTI, Centre for Social and Economic Research on the Global Environment (CSERGE), HMSO, September 1992.

10 *The Economic Value of Externalities from Electricity Sources*, David Pearce, 1992, mimeo.

11 *Regulating the Electricity Supply Industry by Valuing Environmental Effects*, Andrew Stirling, Futures, December 1992.

12 *External costs of electricity generation*, Rainer Friedrich and Alfred Voss, Energy Policy, February 1993.

13 Some of the pros and cons of economic instruments and regulatory approaches are addressed in Making Markets Work for the Environment, Department of the Environment, HMSO, 1993.

14 The Macroeconomic Consequences of Controlling Greenhouse Gases: A Survey, Gianna Boero, Rosemary Clarke and L Alan Winters, *Department of the Environment Environmental Economics Research Series*, HMSO, 1991.

15 *Energy Prices, Taxes and Carbon Dioxide Emissions,* Peter Hoeller and Markku Wallin, OECD Economics and Statistics Department Working Paper No 106, 1991. See Table 4.

CHAPTER 7

Regulatory policies and energy prices[1]

George Yarrow

Introduction

Energy prices are a much favoured target of public policy, as any glance at the history of the energy sector in Britain quickly shows. Traditionally coal has been the major primary fuel used in the economy, and until the 1980's the domestic industry was protected by (i) direct financial assistance to the publicly owned British Coal Corporation (BCC) and (ii) support for both the BCC's volume of sales and its prices in the domestic market. Instruments for price and volume support included guaranteed sales to the state-owned electricity industry at favourable prices and a tax on heavy fuel oil to discourage substitution out of coal. These arrangements all had upward effects on electricity prices. In addition, petrol was subject to a substantial excise tax, motivated more by the revenue requirements of central government than by specific microeconomic objectives.

Today's picture is similar in the sense that there exists a patchwork of different interventions. It is true that government support for the coal industry has been relatively swiftly reduced, but electricity and gas are subject to price controls in franchised monopoly markets - so as to prevent abuse of market power by suppliers - and in England and Wales a levy is applied to electricity generated from fossil-fuel generating sets, the proceeds of which are passed over to the operators of nuclear power stations.[2] While it is possible to justify the levy on environmental grounds (as a precursor to, say, a better-targeted carbon tax), its underlying objective is to provide additional finance for the domestic nuclear industry.[3] Also, until 1994 electricity and gas supplied to domestic premises were not subject to Value Added Tax (VAT), largely due to concern for the position of poorer households whose expenditure on energy tends to be a relatively high proportion of their income. Thus, it is the living standards of the poor that tend to be most adversely affected by the *direct* effects of energy price increases.[4]

Almost inevitably, given the variety of public policy instruments used, some of the interventions mentioned above have had conflicting effects which serve to complicate further the overall energy policy picture. For example, at the same time as attempting to keep prices of electricity to domestic users low by not levying VAT, successive governments have exerted upward pressure on electricity prices by protecting the domestic coal industry.

In part, these conflicts among policy interventions and their effects can be attributed to the multiplicity of government policy aims. For example, government actions may be concerned to: protect a particular producer group from competition; protect consumers from abuses of market power; raise revenues for the Exchequer; redistribute income from richer to poorer households; enhance national security; and promote competition.[5] And, depending upon circumstances, measures taken with one objective in mind can easily have negative consequences for the achievement of other objectives.

It is within this general context of diversity of ends and means that the objectives and instruments of environmental policy are now coming to play an increasingly important role. Discussion of issues such as the effectiveness of environmental regulation, its interaction with other aspects of public policy, and its impact on energy prices need, therefore, to take account of this wider setting. Furthermore, a necessary condition for realistic discussion of regulatory issues in the energy sector is a prior understanding of why policy interventions tend to occur in higgledy-piggledy ways. The next section of the paper will therefore focus on this question, with the aim of establishing a basis for the later discussions of two particular issues in energy regulation: (i) the difficulties in implementing recent proposals for a carbon/energy tax and (ii) potential conflicts between environmental and monopoly regulation.

The importance of distributional effects

Economic analyses of the impact of public policy tend to stress two aspects: effects on economic efficiency and effects on the distribution of income or resources. Much of the implicit framework of academic policy analysis in economics is 'advisory', in the sense that the aim is to assess different policies and to formulate advice on possible improvements. Since it is much easier to achieve consensus on efficiency matters than on distributional matters, the analysis developed by professional economists tends to be biased toward an emphasis on issues of efficiency. One of the most extreme forms of this bias appears in those cost-benefit studies that stress only the final totals of the estimated monetary costs and benefits of a policy, and which therefore set aside problems connected with the distribution of costs and benefits.

Table 1
Public policies and their redistributive implications.

Type of policy:	*Distributive effect:*
National security in respect to sources of fuel	Affects distribution of power among nation states and alliances, and may also favour domestic producer groups
Preventing abuse of market power	Redistributes income between shareholders and customers of the dominant firm, and may also affect the shareholders of rival firms
Tax treatment of domestic fuels	Redistributes income/resources in favour of poorer households
Postalised prices	Redistributes income from consumers at low cost locations to consumers at high cost locations

Regulation in practice

In practice it is very frequently the distributional effects that determine which of two or more conflicting policies are to be adopted, creating a tension between political economy and economic theory. To illustrate, it is not easy to explain the patchwork of policy interventions listed in the introduction by the hypothesis that the interventions simply reflect government attempts to maximise economic efficiency. If, however, the focus is shifted to distribution, it can be seen that many of the objectives and interventions are strongly associated with different types of redistributive effect (see Table 1). The way in which analyses of regulation can be developed from this observation should be fairly clear: rather than redistributive effects being incidental consequences of a drive toward economic efficiency, redistribution of economic resources should be seen as a chief purpose of regulation.

The tension between economic theory and political economy is artificial, however, in the sense that it arises from professional norms rather than anything intrinsic to the intellectual content of theory itself. Indeed, some of the most basic propositions of economics point directly to the likely practical importance of distributional issues. These include the distinction between income and substitution effects and the optimisation principle which underlies most economic analysis.

Income and substitution effects

When the price of a product rises its effects can be decomposed into two components. First, the product will become more expensive relative to other goods, tending to lead consumers to switch (at least to some extent) into what have become relatively cheaper alternatives. This is the substitution effect. Second, the fact that the price has risen while other factors, including money income, have stayed constant means that the consumer will be worse off than before (it will cost the consumer more to purchase a basket of goods that contains some of the product whose price has risen). The amount by which the consumer is worse off can be measured by the fall in real income that would have produced a similar deterioration in the standard of living in the event that prices had not changed. This is the income effect.

The simple message of the distinction between the income and substitution effects is that when market prices change there will not only be effects on behavioural incentives (because some things become relatively cheaper and others become relatively dearer) but also there will be a shift in the distribution of economic resources. Other things being equal, when the price of a product rises consumers will be made worse off and producers will be made better off.[6] Similarly, when a market price falls consumers will be better off and producers will be worse off. Furthermore, the significance of the redistributions will depend upon the relative importance of the expenditure or income to the household or firms concerned. If a household spends a large proportion of its income on a particular product then a change in price will tend to have a significant impact on the wellbeing of that household. Similarly, if a firm depends for most of its revenues on a particular product (i.e. it is relatively undiversified) then its profits will tend to be highly sensitive to the price of the product.

This last point leads to one of the most basic working hypotheses in the economics of regulation: other things equal, policies that have large impacts on each of a small number of households/firms tend to generate larger political payoffs (whether positive or negative) than policies that have small impacts on each of a large number of households/firms. Put simply, those affected in a substantial way tend to notice the effect (it is more visible) and are likely to respond in some way (e.g. by switching their political support); those affected in only a minor way are less likely to notice the consequences of the policy and are less likely to respond. This may be termed the principle of asymmetric response.

When applying this principle it needs to be remembered, of course, that other things may not be equal. Other factors may affect both the visibility of the redistributive consequences of regulatory policies and the responses of individuals and firms to those policies. For example, it is quite possible that, if a policy produces large benefits for a small number of already well-off people, and if those benefits are highly visible to all (i.e. not just the beneficiaries themselves), then the policy may generate negative political payoffs either because of its perceived unfairness or because of the existence of envy. The public response to the salary increases enjoyed by the chief executives of newly privatised utilities may be a case in point.

Consequences of the optimisation principle

Turning to the link between the optimisation principle and the importance of distributional considerations in policy making, consider by way of illustration the problem of setting a tax on an industrial discharge/emission into the environment. Optimality is here defined in terms of maximisation of efficiency in resource allocation, which requires the tax to be set at a level that will lead to the equalisation of marginal costs of abatement and marginal environmental damages. In practice, given the informational problems involved (particularly when measuring damages), this is an extremely complex task, but these difficulties will be ignored for the moment.

Figure 1

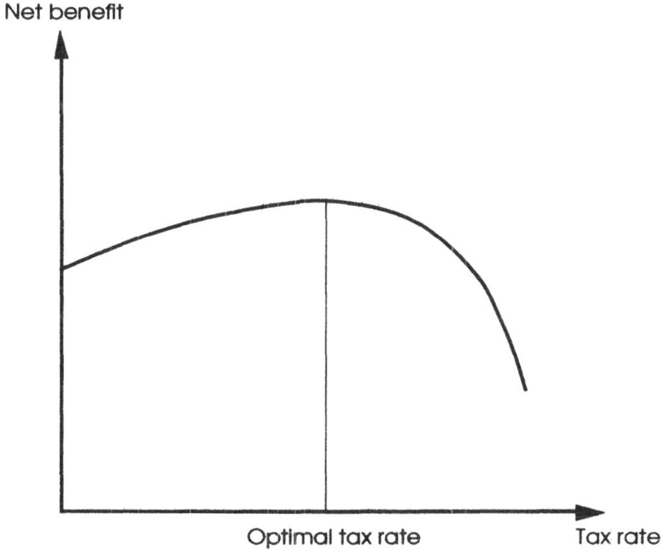

The relationship between economic efficiency - which can be measured as the gain in net social benefits relative to a situation in which environmental discharges/emissions are unregulated - and the tax rate is illustrated in Figure 1. Efficiency first rises with the tax rate and then falls, with the maximum point of the relationship corresponding to the position where marginal abatement costs are equal to marginal environmental damages. This defines the optimal rate of taxation.

It should be noted, however, that small changes in the tax rate in the neighbourhood of the optimal tax will have only very small effects on the level of economic efficiency. The reason for this is simple: provided that the efficiency/tax relationship is smooth - that is, it has an inverted-U shape and not an inverted-V shape - the rate of change of efficiency with respect to the tax must be zero at the efficient tax rate.

The impact of tax changes on the wellbeing of interested parties will, however, be of much greater magnitude around the optimal tax point. Thus, for example, an increase in a tax rate of, say, 10% may have very little effect on economic efficiency, but quite substantial effects on the shareholders and customers of the polluting firms, on those negatively affected by pollution, and, last but not least, on the tax revenues of the government. This gives us a second basic hypothesis: redistribution of resources via changes in regulatory instruments tends to have only very small efficiency costs in the neighbourhood of optimality. This can be called the principle of costless redistribution. It is set out more formally in the appendix.

The principles combined

If we combine the 'asymmetric response' and 'costless redistribution' principles, the result is a set of political incentives that give rise to a persistent tendency to move away from economically efficient outcomes. In terms of political economy, efficient resource allocation tends to be an unstable equilibrium, vulnerable to disturbance by one or other of the interest groups affected by state actions. This is not, of course, to imply that deviations from optimality will grow persistently, but only that the economic system will tend to be locally unstable at an efficient allocation of resources.

A British example: Extension of the VAT base

The case of the imposition of VAT on household fuel bills in Britain in 1994 well illustrates some of the above propositions. Before 1994 there was no VAT on domestic fuels, although business energy use was taxed at the normal rate (17.5% in 1993). The rationale for this position was clearly distributional: energy prices had larger impacts on low income households in general, and poorer elderly people in particular, than on the average member of the population. On the basis of efficiency criteria, stimulating domestic energy usage by reducing the price of fuels relative to the prices of commodities subject to VAT - which was a consequence of the pre-1994 tax regime - could be expected to have negative effects. The pre-1994 policy can not, therefore, be explained by the hypothesis that it was designed to promote economic efficiency.

The proximate cause of the extension of the tax base was a deteriorating position in Britain's public finances (i.e. a widening gap between overall public expenditure and tax revenues). Secondary arguments in favour of the extension of VAT were put forward by the government at the time, including arguments that there would be efficiency gains from increasing incentives to conserve energy. Although well founded, these efficiency arguments appeared to be heavily discounted by the main protagonists in the debate, and the political response to the announcement of the VAT extension was, as might have been expected, dominated by distributional issues.

As a consequence of the hostile public reception given to the announcement (in 1993) of the extension of VAT to domestic fuels - hostility that focused chiefly on the likely effects on the elderly and the poor - the government was forced to introduce additional measures designed to provide some measure of compensation to those most seriously affected. In this way the negative political effects of the policy on the government's level of popular support were mitigated somewhat, albeit at a significant cost to the Exchequer in terms of extra public expenditure commitments.

In this case, then, the disturbance to the initial, non-efficient equilibrium was not caused by a sudden conversion of the government to the cause of promoting efficiency in resource allocation, including efficient use of the physical environment. Rather, a growing crisis in the public finances left one of the relevant parties (the government) dissatisfied with the existing interest-group equilibrium. The government therefore set out to shift the equilibrium but, in the event, found it sensible to settle for a lesser redistribution of resources than was originally intended. Although efficiency was arguably increased as a consequence of the extension of the VAT base, this was largely an accidental consequence of the distributional forces at work. Had it been the case that extra government revenues could only have been raised by increasing inefficiency - as was probably the case in other sectors of the economy at the time - it is highly unlikely that this would have stopped the government raising the level of taxation.

Carbon taxes and global warming

The environmental issues that threaten to have the largest impact on energy pricing are those connected with CO_2 emissions and the prospects for global warming. Full treatment of these issues requires that a number of problems be addressed, including radical uncertainties concerning the measurement of the effects of emissions, the dynamics of climate change and a variety of international jurisdictional problems. We can, however, proceed sequentially, first looking at some estimates of the costs of emissions (and of efficient levels of carbon taxes) in order to establish some initial orders of magnitude.

Estimated effects

Based upon a range of existing estimates of global warming effects, Maddison (this volume) assumes a damage function of the form:

$$D_t = 0.00334W_t^{1.3},$$

where D_t is the fractional reduction in GDP associated with a global temperature increase of W_t. In preliminary runs of his model he estimates that 'business as usual' (i.e. no active steps to reduce project emissions levels) will lead to a temperature rise (from pre-industrialization levels) of 2.15° by 2050 and 3.72° by 2100.

On this basis the percentage reduction in GDP at different levels of warming would be as follows:

Temperature rise	1°	2°	3°	4°	5°
% fall in GDP	0.3	0.8	1.4	2.0	2.7

Translated into $ billions, the impact of a 2° increase in temperature by, say, 2045 gives rise to damage estimates that are very large when compared, say, with the revenue of any one of the UK's leading companies. Nevertheless, when measured relative to the level of world economic activity as a whole, the estimated effect is not a large one.

To see this, consider an index of GDP per head that stands at 100 in 1995 and assume that there is a prospect of 1% growth over the subsequent 50 years. Then GDP per capita in 2045 will stand at around 164.9. A 2° rise in temperature would reduce this to 163.6. That is, the effect of such global warming would be to reduce the prospects for growth in average living standards, measured in terms of per capita GDP, from an increase of 64.9% over a fifty year period to an increase of 63.6% over the same period. Put another way, the warming effect is equivalent to a reduction in the annual growth rate from 1% per annum to 0.984% per annum. Even given a much worse scenario involving a 5° rise in temperature the implied consequence is a reduction in growth prospects over the period from 64.9% to 60.4%, or, in annual terms, from 1% to 0.945% per annum.

It can be seen then that, even if environmental regulations were put in place that led to measures that first stopped and then reversed global warming, the average per capita benefit, *even before deduction of any costs of the measures*, would be extremely modest. When account is taken of the inevitable costs of any measures that are implemented, net benefits can be expected to be lower still. Moreover, halting and reversing global warming represents a relatively extreme form of environmental policy, at least relative to the policy options currently under discussion. Assuming that it could be identified, the optimal or efficient policy is likely to involve measures which have much smaller effects on temperatures. Thus, Maddison estimates that the optimal environmental policy would reduce temperatures by about 0.3^o in 2050 and 0.77^o in 2100, with the effect of increasing GDP (before deduction of costs) by 0.16% and 0.48% respectively in the two years.

On the other hand, in order to achieve these modest changes in GDP in a relatively distant future, current estimates suggest that rather substantial regulatory measures will have to be taken, and taken quickly. These will have effects on energy prices that vary with the form of environmental regulation (command and control, carbon taxes, marketable permits, etc.) but in all cases the magnitudes can be expected to be substantial. To illustrate, the European Commission has proposed a carbon/energy tax at a rate that would rise to the equivalent of $10 per barrel of oil equivalent by the year 2000. This corresponds to an increase of over 70% in the crude oil price as of early 1994. In evidence to the House of Commons Select Committee on Energy (1993), the UK electricity generating company Powergen estimated that this would increase the costs of electricity by about 1.3p per kWh for coal stations and 0.7p per kWh for combined cycle gas turbine stations, assuming an exchange rate of $1.7 = £1. (At early 1994 exchange rates - around $1.5 = £1 - the effect would be somewhat greater.) Taking the overall price effect as the average of these two figures (i.e. 1p per kWh), the effect is to raise domestic prices by around 13% and the prices paid by very large industrial users by around 30% relative to a 1992 base.

More generally, allowing for the various feed-throughs of the carbon/energy tax into prices - via increased petrol prices, higher prices for transportation, higher finished goods prices resulting from higher industrial costs, etc. - the resulting income effects on economic agents will in very many cases be at least one or more orders of magnitude greater than the efficiency effects. Given that compensatory benefits - for example, via cuts in other tax rates - will inevitably be far from perfectly correlated with the burdens imposed by a carbon tax, the predictable consequence is that there will be strong opposition to the tax from interests who can see that their own positions will deteriorate. In other words, as elsewhere in the economics of regulation, distributional issues can be expected to dominate the political economy of carbon taxes.

The international dimension

This last conclusion is strongly reinforced when the international nature of the global warming problem is taken into account. Ignoring for a moment distributional issues within a particular country, the policy problem is exacerbated by the fact that only a fraction of the benefits produced by abatement measures taken in a given country will be captured by the population of that country. And, in the case of a small economy such as Britain, that fraction will be a small one indeed (or, put another way, economic externalities will be large).

Global warming is just one of a wider class of environmental problems in which damages are spread over many countries, and in which the development of regulatory policies is hindered by jurisdictional issues. Where pollution is contained within national boundaries, appropriate actions can be taken by the relevant national government. Where the problem is more localised still it may be dealt with at local government level if the relevant bodies have authority to take the necessary steps to internalise the economic externality. That is, if both polluters and those adversely affected by the pollution are within the same jurisdiction, the controlling authority will be responsible to both sets of parties and should, in theory, take account of both sets of interests in its decisions. (In practice, of course, the influence of the different interests may well differ radically.)

International externalities, then, are just another example of distributional issues in environmental regulation. While cost-benefit analysis proceeds on the basis that the sum totals of monetary costs and benefits should be compared with each other, in practice any change in regulation will, as already stressed, lead to a shift in the distribution of resources. Gainers and losers from any change will often be different groups, and opposition to the change can be expected from the latter irrespective of whether or not there is a net gain in economic welfare as a whole.

The distributional consequences are particularly acute at the international level as a consequence of the wide variations in living standards across the globe. The value of increased environmental quality - and hence the monetary damage caused by any degradation in quality - tends to be an increasing function of income. Other things being equal, a richer country will tend to seek to impose stricter environmental controls than will a poorer country. It is therefore possible that stricter controls that improve resource allocation at the global level would actually reduce economic welfare in less developed nations (i.e. it may lead to energy prices in a low income country that are too high from the point of view of its own, national economic welfare).

It is almost inevitable, therefore, that future progress toward improving the efficiency of environmental regulation will depend upon the development of packages of compensatory payments which serve to improve the position of those most adversely affected by regulatory reform. These, in turn, will depend upon the outcomes of complex, multilateral bargaining at the international level.

The effects of uncertainty

The precise relationships between emissions and damages are, of course, highly complex, and they involve a number of major uncertainties. First, there is uncertainty over how much emitted carbon dioxide stays in the atmosphere and how much is absorbed by the oceans. Second, there are other greenhouse gases that can affect temperatures: CFCs account for a much smaller proportion of greenhouse gas volume but have substantially stronger effects than CO_2. Third, existing atmospheric models do not yet encompass all possible feedback effects (i.e. responses of other variables). Thus, there can be effects on cloud formation, surface ice volumes, ocean circulation, etc. which could mitigate or exacerbate direct greenhouse effects. Fourth, there are significant and uncertain time lags between any increases in gas concentrations and the full emergence of the warming effect. Finally, there is the difficult task of attempting to place monetary values on the various effects.

Since estimates of the relationship between economic efficiency and the relevant tax rate (here levied on the carbon content of fuels) depend upon all of the above exercises, as well as on estimates of the likely economic costs that will be incurred in order to reduce carbon dioxide emissions, it is clear that there will be radical uncertainty about the position and shape of the functional relationship. Hence there will be considerable uncertainty surrounding the appropriate rate at which to set the carbon tax.

In such circumstances, it is highly unlikely that there will be a consensus about how the information that does exist is to be interpreted. Interests tend to influence beliefs, and those with most to lose from higher energy prices can be expected to argue that the environmental problems are being overstated. It does not follow from these points about uncertainty, however, that there will inevitably be a bias against more stringent regulation. Given the nature of the problem, it can be argued that policy should be strongly influenced by 'worst case' evaluations: in particular, environmental policy may be driven by a concern to provide some insurance against highly unfavourable contingencies (although this may simply resurrect similar lines of argumentation concerning the most efficient level of insurance to take out). This is sometimes called the precautionary principle.

Central estimates of the likely monetary effects of CO_2 emissions may not, therefore, be the most important measures in the analysis. Moreover, given lack of knowledge about the precise structure and parameters of the major ecological subsystems (and hence about their stability properties), 'worst case' estimates almost invariably imply significant deviations from a 'business as usual' approach. In other words, the precautionary principle approach might be sufficient to gain general acceptance for much tighter restrictions on emissions and, as a consequence, for significantly higher energy prices.

Where uncertainty may prove to be a more powerful factor in determining the future course of environmental regulation is at the microeconomic level: there exists considerable uncertainty as to the precise spatial patterns of the effects of global warming. The regional information necessary to predict gainers and losers from uncompensated regulation, and, more importantly, the magnitudes of the likely gains and losses, is still extremely thin. As a consequence, the international bargaining process is complicated by participants playing information games designed to buttress their own national interests.

Opposition to carbon/energy taxes

In view of the above points, it is not surprising that proposals to implement measures as far-reaching in their consequences as carbon and energy taxes have met with fierce opposition. In the USA, for example, President Clinton's initial proposals for a substantial energy tax were almost entirely whittled away by Congress. Implementation of the proposed European carbon/energy tax was made conditional on, among other things, (i) the introduction by other OECD countries of similar measures that have an equivalent financial impact, and (ii) the introduction by Member States of measures to ensure that the effect on the overall tax burden in each State is neutral. As is to be expected, energy intensive industries in Britain and Europe have been among the interest groups that have led the opposition to the carbon/energy tax, fearing the imposition of measures that would increase costs in the EC relative to competitors elsewhere in the world. If the tax was being levied so as to produce benefits that accrued entirely within the UK, say, it would in theory be possible for the government to provide financial compensation to energy intensive industries (so as to protect their competitive position) while at the same time delivering a net improvement in UK economic welfare. Such an outcome is, however, not possible in the global warming case precisely because widespread international externalities mean that unilateral actions by the UK (or the EC) would not lead to a net improvement in national economic welfare. That is, the domestic benefits would be insufficient for the government to raid them in order to raise finance for compensatory payments to industry.

What this means is that local (i.e. within a given jurisdiction) distributional issues and global distributional issues must be considered together if the obstacles to improving the efficiency of regulation are to be overcome. Needless to say, it is a formidably difficult task to resolve all the various issues together, and distributional problems can therefore be expected to continue to act as a severe constraining influence on what can be achieved in the regulatory field.

Potential conflicts between monopoly regulation and environmental regulation

It is perhaps ironic that, in Britain, the gradual tightening of constraints on emissions of waste gases from fuel combustion has coincided with a period in which new regulatory regimes for the privatized gas and electricity industries have been developed. For it is immediately apparent that there is potentially a major conflict of philosophies between those parts of public policy that are concerned with problems of monopoly and competition and those parts that are concerned with the environmental effects of energy production. Put in a nutshell, the traditional concern of monopoly regulation is that outputs are inefficiently low, and that prices are inefficiently high, as a consequence of the exercise of monopoly power by the major utilities. On the other hand, the concern from the environmental side is that outputs may be inefficiently high (as well as being produced by excessively polluting technologies). This is well illustrated by the ways in which the two problems are handled in standard expositions of economic theory, which can be summarised as follows:

Monopoly. A profit-seeking firm with market power chooses an output where marginal revenue equals marginal cost. Since marginal revenue is below price (the demand curve is downward sloping) it follows that price exceeds marginal cost. The value of an extra unit of output (price) therefore exceeds the cost of producing it, which implies that output is lower than is optimal (or, in the alternative formulation, that price is too high).

Environment. In maximising profit the firm takes no account of environmental damages. The firm equates marginal revenue with marginal private costs rather than with marginal social costs. Since social costs are higher than private costs, efficiency would be improved if output were lowered.

It is important to note at this point, that introductory treatments of environmental costs tend to work within the model of an otherwise competitive market (marginal revenue is assumed equal to the market price of output). In this case, the finding that marginal social costs exceed marginal private costs leads directly to the conclusion that output is too low. Similarly, when examining monopoly and deciding whether the outcome is efficient or not, the benchmark for comparison is the 'competitive' equilibrium characterised by price equals marginal cost. In this way economic problems are, in effect, considered one at a time, assuming that all other conditions for optimality in resource allocation are satisfied. In other words, in both cases we are *within one policy change of a first-best (efficient) allocation of resources*. However, while this may be a useful way of introducing economic concepts and issue, it is not a sensible way of analyzing regulatory policies: policy analysis is invariably located in a world that is much further away from the first-best. Suppose, therefore, that instead we consider both issues together, within in a common framework. What can then be concluded?

Targets and instruments

Suppose initially that the utility is required to pay an environmental 'tax' per unit of its output on the assumption, which will be relaxed later, that there is a fixed relationship between its output and the level of its waste gas emissions (i.e. we are looking at an energy tax, rather than a carbon tax). The firm will build this tax into its private cost function and charge a markup on marginal cost which depends upon demand conditions.

In this case, if the tax is set so as to reflect the best estimate of the marginal environmental damage caused by an additional unit of output, then the monopoly markup will ensure that price is above the socially efficient price and output is below the socially efficient output. There is, however, no reason to assume that the environmental tax will be set in this way, for now we are dealing with a second-best world where the tax instrument should be used in a way that takes account of predictable distortions elsewhere.

In fact the environmental charge should be a sufficient instrument for achieving the socially efficient price and output levels on the market. To take the simplest case, let both marginal cost (c) and the elasticity of demand be constants. The utility's markup over marginal cost is then also a constant, say m, and market price, p, will be:

(1) $$p = m\,[c + t],$$

where t is the environmental charge. Hence, any given market price, including the optimal price, can be achieved by setting:

(2) $$t = [p/m] - c.$$

Note that in some cases the optimal tax rate t may be negative, in which case output is subsidised. This will tend to occur when environmental damages are relatively low (i.e. p is not much greater than c) and monopoly power is high (i.e. the markup m takes a high value). The result here is fairly intuitive: when the monopoly problem is severe the market output will be too low, even allowing for environmental costs of extra output, and subsidy is one means of inducing greater production.

Under the given, simplified conditions, one policy instrument - a tax or subsidy on output - is sufficient to induce the efficient equilibrium. However, in dealing with the efficiency issue, the use of only one policy instrument means that the government is unable simultaneously to control the distributive effects of intervention.

The positions of the firm and the government are as follows

(3) supernormal profit $= [p - c - t]q$

$= pq - pq/m$, from equation 2 above.

(4) government revenue $= tq$

$= pq/m - cq$. also from equation 2.

Thus, if the tax rate is set so as to achieve efficiency in resource allocation, these financial flows are fully determined.

Multiple instruments

If, now, we add a second policy instrument, the picture changes somewhat. One possibility would be a lump sum tax on the firm equal to the value of monopoly profits. This would be like public ownership in that all profit would go to the state. The other obvious possibility is for the government to control final selling prices, which would give control of the markup m and, hence, of both the profit and tax/subsidy flows. Thus, as can be seen from the above expressions, as the markup m decreases, the share of total surplus appropriated by the state increases and the share of the surplus appropriated by the firm decreases. Here there are two objectives - efficiency and distribution - and two instruments to achieve these objectives, the environmental tax and utility price controls.

Alternatively, to the extent that it can be used as a policy instrument, affecting the 'degree of competition' it can also be a means of shifting the distribution of resources. For example, the promotion of competition in energy markets would be one way of seeking to achieve an efficient equilibrium in which monopoly rents are transformed into government revenues.[7] Thus, the more competitive the market the higher will be the optimal level of the environmental tax.

Similar arguments apply in cases where there is not a fixed relationship between outputs and emissions. In this case it is appropriate for environmental regulation to seek to tax or control the level of emissions (rather than output), perhaps by using some proxy such as the carbon content of the fuels that are burned. The difference now is that two policy instruments are required to achieve an efficient equilibrium, since the latter requires both an efficient choice of technologies for a given level of output and an efficient output level. However, with two instruments - say a carbon tax and an output tax/subsidy - regulatory policy still has no degrees of freedom with which to meet distributional objectives. Thus, in this case, the pursuit of distributional objectives will again tend to involve trade-offs with economic efficiency.

Regulatory sprawl

What the above analysis shows is that, within an overall policy framework that includes environmental charges or other environmental instruments, there is nothing inherently problematic about utility price capping that encourages the firm to increase its output, or with policies of promoting competition that also have the effect of increasing output. The reason for this is simply that such tendencies can be offset by the other instruments: the more aggressive the output policy of the firms the tighter are the environmental controls required to achieve economic efficiency. Moreover, from a governmental perspective, price-capping or competition has the advantage of giving rise to an efficient equilibrium in which government revenues are higher.

Nevertheless, the incentives for output expansion generated by RPI-X price control formulae have been criticised on the ground that they encourage activity that leads to increased environmental damage. As a consequence, there has been pressure on regulatory bodies set up to deal with issues of monopoly and competition to extend their interests into the field of environmental regulation.

One way in which these pressures have been accommodated is via movement away from capping prices or average revenues and toward capping total revenues. The argument is as follows: if the regulated utility's maximum allowed total revenue is fixed according to an RPI-X+Y formula, the firm will no longer be able to increase its revenue by expanding output. Indeed, since marginal revenue is always zero, it will be below marginal cost. In consequence the firm will have incentives to *reduce* output.

A total revenue constraint is an extreme option, but alternatives include controls that are combinations of caps on total and average revenues. For example, total revenue could be constrained to lie below $F + p^*q$, implying that:

$$pq < F + p^*q.$$

Under this system the price control operates like a two part tariff such that the incremental revenue to the firm is effectively p^*. Arguably, p^* should be set equal to marginal cost for first-best incentives.

While such adjustments to price control formulae are sensible enough when they reflect the underlying structure of costs - so that, for example, total revenue caps are associated with fixed costs and average revenue caps with variable costs - arguments in their favour based on environmental effects should be treated with some scepticism. In earlier sections of the paper, it was argued that efficient regulation is highly vulnerable to interest group pressures, and such pressures can be applied to any institution able to 'supply' regulation. Thus, provided that they are not totally constrained by their legal frameworks, institutions set up to achieve one set of regulatory goals can often be induced to take on board other objectives, particularly if the additional objectives can be achieved without great loss of efficiency.[8]

Where this happens the likely outcome is what may be termed 'regulatory sprawl'. The activities of different agencies start to overlap and objectives become more confused. Co-ordination of regulatory activities becomes increasingly complex, and evaluation of regulatory performance is made more difficult. In the longer term, the consequent lack of transparency in the regulatory system is likely to make it even more susceptible to influence by interest groups seeking measures that secure a favourable redistribution of economic resources.

Conclusions

As with other forms of economic regulation, environmental regulation is heavily influenced by conflicts over the distribution of economic resources, and hence efficient outcomes are highly vulnerable to distributional pressures. It follows that the effective conduct of public policy in this area requires much more than just greater scientific understanding of the environmental effects of economic activity coupled with cost-benefit analysis and the application of market based control mechanisms (environmental taxes, marketable permits, etc.) Challenging problems of political economy lie ahead.

The importance of compensatory payments in facilitating tax changes that lead to higher energy prices was well demonstrated by the case of the extension of VAT to domestic fuels in Britain. When dealing with issues of the scale of global warming, however, the distributional problems are orders of magnitude more difficult to handle, involving multilateral negotiations over measures (or the absence of measures) that have highly uncertain, but potentially very far reaching, consequences. Political skills will be at a premium in the task of preventing narrow self interests leading to badly suboptimal global outcomes.

Over the past years the increasing stringency of environmental controls on the energy industries has inevitably had an upward impact on energy prices, and this is a trend that can certainly be expected to continue, if not accelerate. This does not mean, however, that there should be major changes in those aspects of public policy toward energy industries that involve price controls and the promotion of competition. It is true that, other things being equal, these policies are aimed at lowering, rather than raising energy prices. However, other things should not be held equal: for example, the more effective monopoly policy is in holding down prices, the higher should be the level of any carbon or energy tax.

In helping to ensure that given environmental targets can be met at less cost, successful monopolies/competition policies can serve to enhance environmental regulation. Such policies are, however, likely to be more effective if they remain focused on the issues they were designed to tackle. Large scale extension of the activities of energy utility regulators into the environmental field is, in contrast, likely to detract from the overall performance of regulatory systems.

Appendix

Let environmental benefits and abatement costs be represented by $B(t)$ and $C(t)$ respectively, where t is the tax rate. Maximisation of economic efficiency (E) can be taken as being equivalent to the maximisation of the difference between benefits and costs, leading to the first-order condition:

$$dE/dt = dB/dt - dC/dt.$$

Suppose, however, that policy makers seek to maximise the objective function:

$$B(t) - (1 + \quad)C(t),$$

where   is a parameter that reflects distributional factors. For example, when   > 0 the regulatory system could be viewed as having a bias towards producers interests - the costs of abatement are given more weight than its benefits - and when   < 0 the regulatory system has a bias towards the interests of those most badly affected by pollution (= 0 corresponds to maximisation of economic efficiency).

The first-order condition for optimality is now:

$$dB/dt - (1 + \quad)dC/dt = 0.$$

Solving this for t, we can express the equilibrium tax rate as a function of the distribution parameter, t( ). The impact of a change in distributional objectives can be determined by totally differentiating with respect to :

$$dt/d \quad = (dC/dt)/[d^2B/dt^2 - (1 + \quad)d^2C/dt^2] < 0,$$

assuming that second-order conditions for a maximum are satisfied (in which case the denominator is negative). That is, as is to be expected, giving extra weight to the interests of producers leads to a lower equilibrium tax rate.

The 'costless distribution' principle follows from differentiation of (maximised) utility with respect to :

$$dE/d \ = (dE/dt)(dt/d\ ) = 0.$$

This equation says that, starting from an efficient equilibrium, there will be no penalty in terms of lost efficiency attaching to a small shift in regulatory objectives in favour of one or other of the interest groups. On the other hand:

$$dB/d \ = (dB/dt)(dt/d\ ) = (dC/dt)(dt/d\ ) = dC/d \ < 0.$$

That is, the impact of the shift in objectives on benefits and costs, considered separately, is not vanishingly small. If benefits and costs accrue to different interest groups, this implies that the distributional effects of the change in objectives are much greater in magnitude than the efficiency effects.

Notes

1 I am grateful to Mike Parker, Mark Schofield and Nick Woodward for comments on an earlier draft of this paper.

2 Similar effects are achieved in Scotland by means of contracts for the sale of the output of Scottish Nuclear,

3 Alternatively, such support could be supplied from general taxation, in which case electricity prices would be lower than is currently the case.

4 Indirect effects of higher energy prices include the increased cost of car journeys, which may have larger impacts on households further up the income scale.

5 Promotion of competition in the electricity and gas industries was an increasingly important objective of UK regulatory policy in the 1980s and early 1990s.

6 Since a price increase will have been triggered by some change in economic circumstances, the "other things equal" qualification is important here. For example, if price increases are triggered by higher taxes it may well be the case that producers, as well as consumers, are worse off at the new equilibrium. On the other hand, *given the tax increase*, producers will be better off if prices are raised than if they are held constant.

7 Competition does, of course, have many other effects, including producing incentives for cost reduction. These are neglected in the current analysis.

8 Moving into new areas tends to increase the power of the regulatory agency, contributing to internal "bureaucratic" objectives.

CHAPTER 8

Economic evaluation of different generating systems, environmental cost and carbon tax simulations

Kazuya Fujime

Economic evaluation of different generating systems

This study makes an economic assessment of different generating systems (nuclear, LNG, coal, oil thermal power). It does this by estimating the generating costs of plants commissioned in the years 1992 and 2000 using current and likely levels of unit construction cost, fuel price, and other conditions.

Estimation method

In this study, the generating cost is estimated in terms of life-long annually equalized cost for the power plants commissioned in the target years. This can be obtained by first converting the life-long cost incurred in newly-built plants into present values, then dividing the outcome by their life-long generated output.

This method is in common use and is considered to provide a fairly even basis for evaluating capital-intensive generating sources at the initial stage, e.g. nuclear, and other fuel cost-dominated sources, such as thermal power.

The main assumptions employed in the estimation are listed below. No estimation is made for the oil-fired thermal power plant commissioned in the year 2000, because no plant construction is planned from 1989 onwards.

(1) Target years: 1992, 2000
(2) Target power resources: Nuclear, LNG, coal, oil (1992 alone)
(3) Price declaration: 1992 price in real terms
(4) Exchange rate: ¥127/$ (kept constant in the future)

The service life specified in Japanese tax law (16 years for nuclear power plants, and 15 years for thermal power plants) was used in making the estimations.

Assumptions for generating cost estimation

Source-by-source unit construction costs of the power plants commissioned in 1992 and those in 2000 were obtained first, by extracting plant-by-plant unit construction costs from the 1992 Generating Capacity Installation Plan, by referring to actual and planned figures for a total of seven years (the three years before and after each base year). Subsequently, resultant unit costs were adjusted by commodity price as well as by unit number; then, by averaging the outcome, unit construction costs were produced for the model plants onstream in 1992 and those in 2000.

The unit construction costs extracted from the Generating Capacity Installation Plan were adjusted by commodity price in order to put them in 1992 prices. Commodity price-based adjustment was made by taking the Bank of Japan's wholesale price index as the base. For the portion up to 1992, per-annum average actual records (an April-December average in the 1992 case) were taken, while a 1.0% rise a year is assumed for the portion from 1993 onwards.

Even if built on the same site and given the same specifications, construction costs of the first unit and subsequent units in general vary considerably. This is because the cost to construct the first unit often includes cost of land, land reclamation, and of installations such as port facilities which can be put to common use with subsequent units. At the same time construction of the second and subsequent units tends to cost less, due to the effects of standardization, acquisition of skills, and a shortened period of land reclamation. These factors must be taken into account in order to produce correctly estimated unit construction costs for the model plants.

In this study, unit construction costs of the model plants are adjusted by taking an average commodity price-adjusted unit construction cost of all the units commissioned in a relatively short period in a single site, based on which coefficients of unit number-based adjustment were calculated for the first unit as well as the second and subsequent units.

Nuclear fuel price scenario

For every stage of the simulation, the evolution of nuclear fuel prices is based on the following assumptions:

- Due to a slowdown in nuclear development in recent years, uranium concentrate supply/demand is slack at present. Accordingly, it is assumed that price remains virtually flat between 1992 and 2000. From 2001 onwards, it is assumed to grow 1.0% annually given likely uranium supply/demand developments and potential reserves.
- Considering that the UF_6 process enters a mature stage in technical terms, the conversion cost is assumed to be flat onwards from 1992.
- The uranium enrichment cost was produced by averaging the prices of enrichment offered by the U.S. Department of Energy and the EURODIF by their shares in enrichment service. Given technical maturity, both their enrichment prices are assumed to virtually level off from 1992 onwards.
- Given technical maturity, the fabrication cost is assumed to remain virtually stable from 1992 onwards.
- From 1992 onwards, the spent fuel transport cost is assumed to remain virtually stable.
- The spent fuel reprocessing price is assumed to remain virtually stable from 2000 onwards.

Fossil fuel price scenario

In the base case, fossil fuel prices are assumed to rise in the long run. For oil, coal and LNG, high and low price cases were also used in the simulations in addition to the base case. Table 1 shows the specific results of fossil fuel price scenarios.

In 1992, the crude oil (CIF) price rose from $18/bbl to $20/bbl where it levelled off. Thus, given that the price has stayed at the $20/bbl level since the second quarter, the crude oil price in 1992 is put at $20/bbl. The future price was assumed, by referring to projections made by the IEA and other oil-related organizations, at $26/bbl for 2000 and $31/bbl for 2010 in the base case, $21 and $26/bbl in the low price case, and $31 and $36/bbl in the high price case.

Table 1
Fossil fuel price scenarios

Oil - Base case

FY	Crude oil CIF ($/bbl)	Imported C heavy fuel oil ($/bbl)	Exchange rate (¥/$)	Domestically refined C heavy fuel oil (¥/$)	Imported C heavy fuel oil (¥/bbl)	Product (¥/kl)	Synthetic Price (¥/kl)	Crude oil burning (¥/kl)	Pre-burner Price (¥/kl)
1992	20.0	18.0	127	24,050	23,474	25,180	24,332	22,201	23,266
1995	23.0	21.0	127	26,502	25,91	27,637	26,783	24,653	25,718
2000	26.0	24.0	127	28,954	28,348	30,094	29,235	27,105	28,170
2005	28.5	26.5	127	30,997	30,379	32,141	31,279	29,148	30,213
2010	31.0	29.0	127	33,041	32,410	34,188	33,322	21,192	32,257

LNG - Base case

FY	LNGCIF ($/t)	Exchange rate (¥/$)	Expenses (¥/t)	Pre-burner price (¥/t)
1992	188.9	127	1,020	25,010
1995	208.1	127	1,020	27,449
2000	227.3	127	1,020	29,887
2005	243.3	127	1,020	31,919
2010	259.3	127	1,020	33,951
2015	275.3	127	1,020	35,983

LNG - High price case

FY	LNGCIF ($/t)	Exchange rate (¥/$)	Expenses (¥/t)	Pre-burner price (¥/t)
1992	188.9	127	1,020	25,010
1995	224.1	127	1,020	29,481
2000	259.3	127	1,020	33,951
2005	275.3	127	1,020	35,983
2010	291.3	127	1,020	38,015
2015	307.3	127	1,020	40,047

LNG - Low price case

FY	LNGCIF ($/t)	Exchange rate (¥/$)	Expenses (¥/t)	Pre-burner price (¥/t)
1992	188.9	127	1,020	25,010
1995	192.0	127	1,020	25,417
2000	195.3	127	1,020	25,823
2005	211.3	127	1,020	27,855
2010	227.3	127	1,020	29,887
2015	243.3	127	1,020	31,919

Coal - Base case

FY	Steaming Coal CIF ($/t)	Exchange rate (¥/$)	Expenses (¥/t)	Pre-burner price (¥/t)
1992	48.5	127	1,700	7,856
1995	50.8	127	1,700	8,157
2000	53.2	127	1,700	8,458
2005	55.2	127	1,700	8,708
2010	57.2	127	1,700	8,959
2015	59.1	127	1,700	9,210

Coal - High price case

FY	Steaming Coal CIF ($/t)	Exchange rate (¥/$)	Expenses (¥/t)	Pre-burner price (¥/t)
1992	48.5	127	1,700	7,856
1995	52.8	127	1,700	8,408
2000	57.2	127	1,700	8,959
2005	59.1	127	1,700	9,210
2010	61.1	127	1,700	9,461
2015	63.1	127	1,700	9,712

Coal - Low price case

FY	Steaming Coal CIF ($/t)	Exchange rate (¥/$)	Expenses (¥/t)	Pre-burner price (¥/t)
1992	48.8	127	1,700	7,856
1995	48.9	127	1,700	7,906
2000	49.3	127	1,700	7,956
2005	51.2	127	1,700	8,207
2010	53.2	127	1,700	8,458
2015	55.2	127	1,700	8,708

By referring to the formula employed by Tokyo Electric and Japan Oil in their 1992 contract, the pre-burner price of oil was obtained by calculating the prices of domestically refined C heavy fuel oil, imported C heavy fuel oil, desulfurized product, and crude oil burning, and taking the weighted average price of the outcome. The formula was assumed to remain unchanged in the future.

The past trend shows that the LNG price is highly correlated to the crude oil price. Accordingly, the correlation function between the crude oil and LNG prices was estimated from the past data and from this the LNG (CIF) price in the future was assumed. The price corresponding to the crude oil price base case was taken as LNG base case, and those prices corresponding to the high and low cases as LNG high and low cases. The pre-burner price of LNG was obtained by adding domestically incurred expenses (¥1,020/t) to the LNG price obtained as described above.

As for the LNG price, the coal price was assumed to be correlated to the crude oil price. The pre-burner price of coal was calculated by adding domestically incurred expenses (¥1,700/t) to the coal CIF price.

Reactor decommission cost

The cost of decommissioning reactor installations was calculated based on the estimation (¥30 billion for a 1.1 GW-class plant) made by the Advisory Committee for Energy's Nuclear Session in its interim report released in 1985. The decommissioning cost was assumed to be spent in a lump sum after the full service life (16 years).

Other assumptions

Other assumptions employed in generating cost estimation are shown in Table 2.

Gross heat efficiency of LNG thermal power generation was put at 44% for plant commissioned in 2000 on the assumption that the combined cycle type, characterized by outstanding heat efficiency, could be introduced by that time.

Table 2: Other assumptions

Depreciation	15 years	(16 years for nuclear)
Rate of residual value	10%	(percentage on diminishing value
Rate on investment	7.2%	
Rate of property tax	1.4%	
In-plant rate	Nuclear LNG Coal Oil	4.0% 3.6% 8.4% 3.5%
Utilization factor	Nuclear LNG Coal Oil	4.0% 3.6% 4.8% 3.9%
Long-term discount rate	6%	
Heat value of fuels	LNG Coal C heavy fuel oil	13,000 kcal/kg 6,200 kcal/kg 9,800 kcal/kg
Gross heat efficiency	LNG Coal Oil	39% (44% when commissioned in 2000) 39% 39%

Estimated results

Estimated results are presented in Table 3. For the plants commissioned in 1992, only the base case of fossil fuel scenario was assumed, while the base, high and low cases of fossil fuel scenario were prepared for those commissioned in 2000. In making the estimation, the plants were assumed to work at 70% of their maximum capacity.

Table 3: Results of generating cost estimations (Equalized cost, gross, utilization factor 70%)

Plants commissioned in 1992 (¥/kWh)

	Nuclear	Thermal		
		LNG	Coal	Oil
Capital cost	6.36	4.22	5.37	3.86
Operating costs	2.32	1.39	2.35	1.38
Fuel cost	1.47	5.03	3.22	6.26
Total	10.14	10.64	10.94	11.51

Plants commissioned in 2000 (¥/kWh)

	Nuclear	Thermal					
		LNG			Coal		
		Low	Base	High	Low	Base	High
Capital cost	5.91		4.31			5.72	
Operating cost	2.15		1.42			2.50	
Fuel cost	1.47	4.40	5.03	5.67	3.2	3.40	3.59
Total	9.52	10.13	10.77	11.40	11.43	11.62	11.82

Breakdown of nuclear fuel cycle cost (¥/kWh)

	Nuclear fuel cost
Uranium concentrate	0.32
UF_6 conversion	0.03
Enrichment	0.35
Fabrication	0.32
SF transport	0.08
Reprocessing	0.37
Total	1.47

Plants commissioned in 1992

Among the plants commissioned in 1992, nuclear power generation proved least expensive; it was costed at ¥10.14/kWh, compared with ¥10.64/kWh of LNG thermal power, ¥10.94/kWh of coal thermal power, and ¥11.51/kWh of oil thermal power. Also, the results showed that LNG was more economical than any other thermal power generating systems.

Focusing on the gaps among the generating sources, the most inexpensive, nuclear, was 12% cheaper than the most expensive, oil thermal power. With oil thermal power excluded, gaps among the alternatives are limited within the range from 5% (LNG thermal power) to 7% (coal thermal power).

The cost to decommission nuclear installations amounted to ¥0.18/kWh. This cost was included in capital cost shown in Table 3.

However, this cost estimation did not allow for the high-level waste disposal cost incurred at the end of the nuclear fuel cycle because specific methods of disposal and cost calculation were both still under consideration, although it had been mapped out as the basic policy that 'high-level wastes be confined in glass, then disposed of deep underground.'

Plants commissioned in 2000

Among the plants commissioned in 2000, nuclear power generation would cost ¥9.52/kWh, LNG thermal power ¥10.77/kWh in the fossil fuel base case, and coal thermal power ¥11.62/kWh. Thus, the gaps between nuclear and thermal power generation would widen to 12-18%. Moreover, in the fossil fuel high price case, nuclear would cost ¥9.52/kWh, compared with ¥11.40/kWh of LNG thermal power and ¥11.80/kWh of coal thermal power. As a result, the already widened gaps would increase, making nuclear 16-19% cheaper than the alternatives.

On the other hand, in the fossil fuel low price case, where fuel prices are likely to rise only slightly until 2000, nuclear would cost ¥9.52/kWh, compared with ¥10.13/kWh of LNG thermal power and ¥11.43/kWh of coal thermal power. Although the gaps between nuclear and thermal power generation would be narrowed to 6-17%, nuclear's economic superiority remains unshaken. This is attributable to the falling generating costs of nuclear, thanks to lower construction costs than required for plant onstream in 1992 plus stable fuel cost. It is also related to the fact that thermal power generating costs would rise due to increasing fossil fuel prices.

These would consequently precipitate diverging economics between the two.

Figure 1: Sensitivity analyses of generating costs

(When commissioned in 1992)

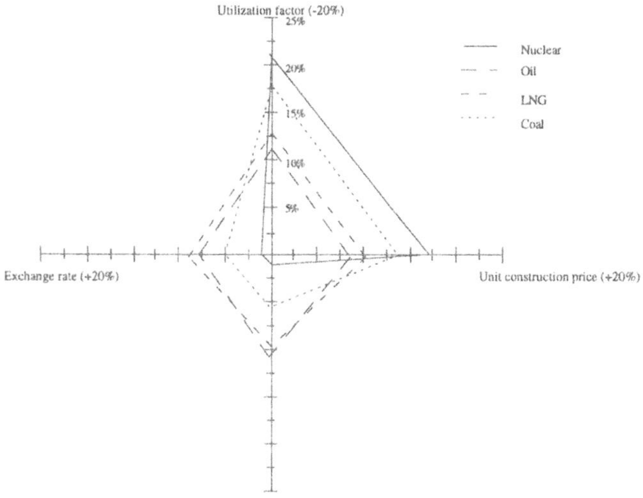

* Up/down in generating cost when these parameters fluctuate by 20% each

(When commissioned in 2000)

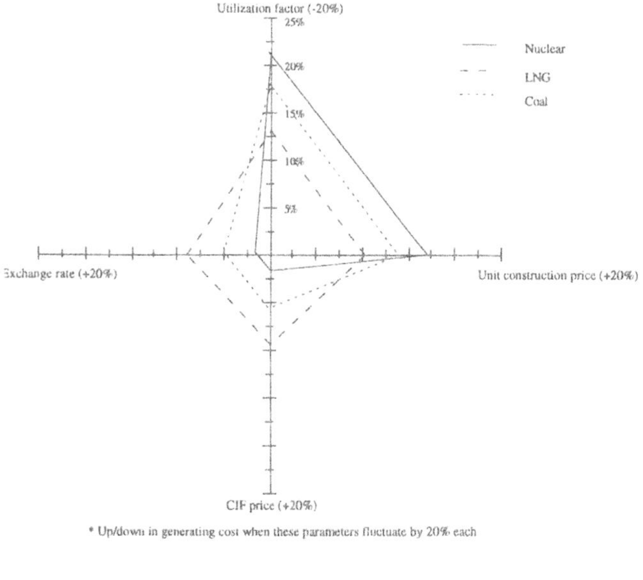

* Up/down in generating cost when these parameters fluctuate by 20% each

Figure 2: Generating cost by utilization factor

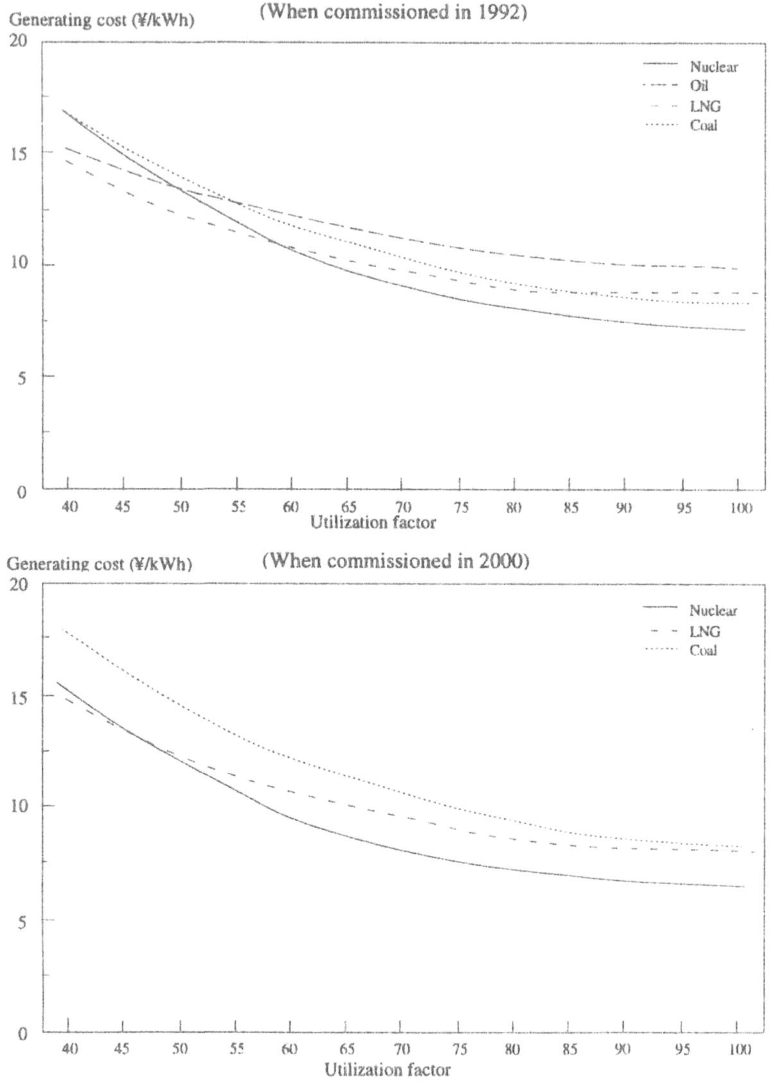

Among thermal power generation, despite rising fuel prices, the fuel cost of LNG thermal power would be flat in the fossil fuel base case, and some 10% lower in the low price case, due to high heat efficiency. In contrast, in reflection of a combination of rising fuel prices and growing construction cost, coal thermal power would result in the ¥11/kWh level in all the three cases.

Figure 3: Generating costs by construction cost

(When commissioned in 1992)

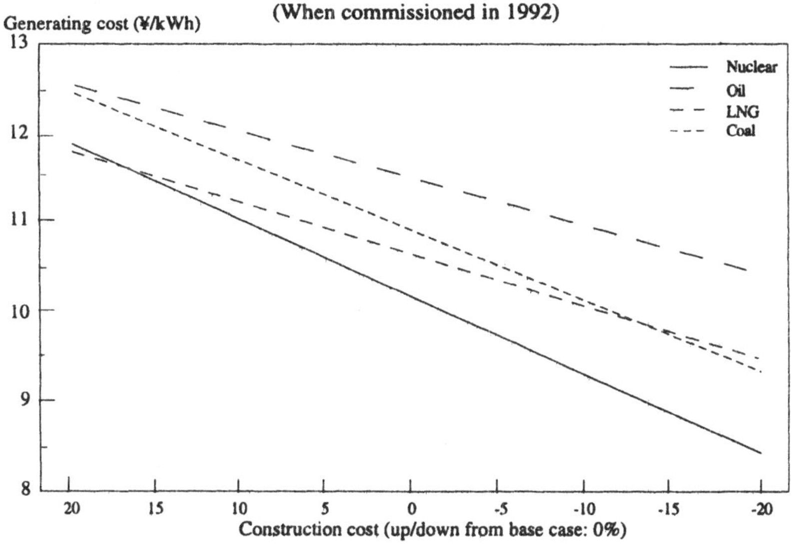

(When commissioned in 2000)

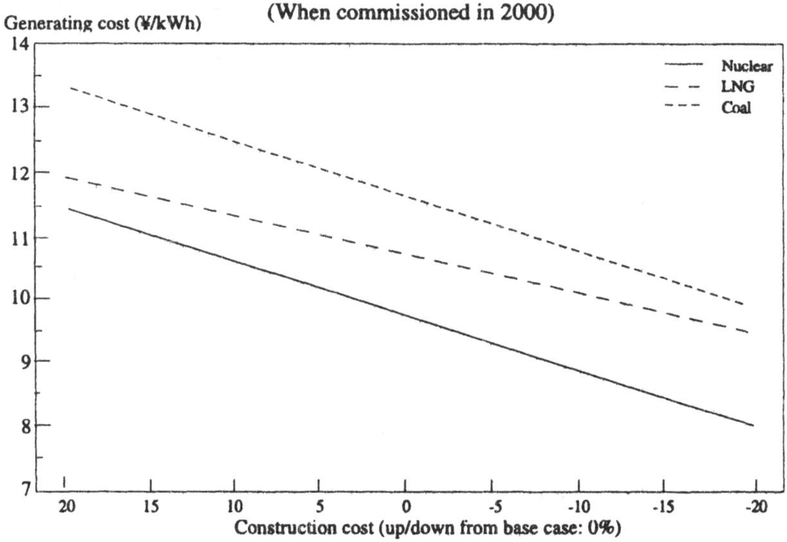

Sensitivity analysis

The generating costs of different generating sources are affected by different factors. Figure 1 illustrates how much the generating costs can increase when the utilization factor, unit construction costs, exchange rate, and pre-burner price of fuel each fluctuates 20% from the base case. The chart shows that nuclear is very sensitive to fluctuations in the utilization factor and unit construction cost, thus confirming that nuclear is 'capital-intensive.' On the other hand, LNG thermal power is severely affected by fluctuations in fuel price, exchange rate and fuel cost, proving that it is 'fuel cost-dominated'.

Utilization factor

Among the plants commissioned in 1992, nuclear plants would be more economical to run than oil thermal power ones when the utilization factor stands at 50%, and LNG thermal power at 60%. With its utilization factor becoming higher, nuclear's economic superiority would increase (Figure 2). Plants onstream in 2000 would show similar trends, though the gap between nuclear and thermal power generation would widen and nuclear would become the least expensive generating source in the domain of a utilization factor higher than 45%.

As discussed above, nuclear is the cheapest among the plants onstream in 1992 when running at 60% or higher, and among those in 2000 when kept at 45% or higher. Moreover, it is shown that the higher utilization factor would result in greater cost effectiveness.

Unit construction cost

Among the target generating sources, 'capital-intensive' nuclear is sensitive to fluctuations in construction cost. Figure 3 shows generating costs resulting when unit construction costs fluctuate from the base case. A fluctuation in generating cost, which could result from a 1% fluctuation in unit construction cost, would amount to a range of ¥0.085/kWh (1992) - ¥0.079/kWh (2000) in the nuclear case, compared with LNG thermal power's ¥0.056 - ¥0.057/kWh range, and coal thermal power's ¥0.077 - ¥0.078/kWh range. When commissioned in 1992, nuclear and LNG thermal power plants could compete with each other if their unit construction costs increase by around 15% each (¥392,000/kW for nuclear, and ¥262,000 for LNG thermal power).

To calculate the effects of unit construction costs, the same assumptions were taken as in the fossil fuel base case, which were applied to the plants commissioned in 1992 and those in 2000.

Cuts in nuclear construction costs are projected in the long run. This fall in unit construction cost would form a crucial factor in promoting the economic superiority of nuclear in the future. However, the projection would not necessarily become reality. Therefore, assuming that the thermal power plants commissioned in 2000 would show no fluctuations in their unit construction cost, calculations were made to learn to what extent unit construction cost of nuclear could rise before being challenged by thermal power generation in economic terms. The results are shown in Figure 4.

Figure 4: Competitive prices between nuclear and thermal power generation (by construction cost)

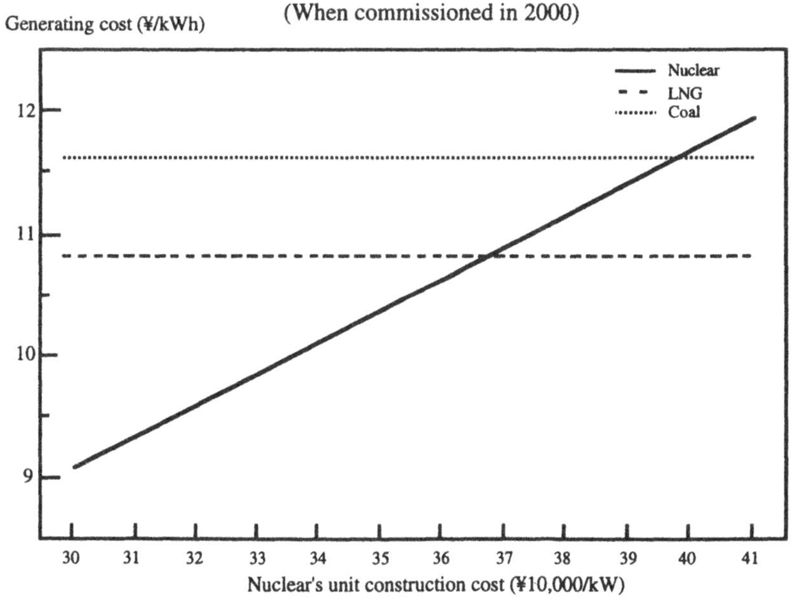

These results show that nuclear would be challenged by LNG thermal power when its unit construction cost stood at ¥366,000/kW, and by coal thermal power at ¥400,000/kW. This means that nuclear would be challenged by LNG thermal power when its unit construction cost increased 7% above the present level (at the plants commissioned in 1992), and 16% above the projected level (for the plants commissioned in 2000).

Likewise, it would be challenged by coal thermal power if the unit cost was 17% and 27% higher than the present and projected levels, respectively.

Exchange rate

The exchange rate, assumed at ¥127/$ when estimating the generating costs, is now floated in the range of ¥107-147/$. It should be noted, however, that each of the resultant exchange rates in this range is assumed to remain constant.

Among the target generating systems, thermal power generation, heavily weighted by fuel cost, is sensitive to exchange rate fluctuations. In a striking contrast, nuclear is affected little by such fluctuations. Every ¥1/$ fluctuation in the exchange rate causes nuclear generating costs to fluctuate as little as ¥0.004/kWh (1992) - ¥0.005/kWh (2000), compared with ¥0.38/kWh for LNG thermal power generating cost, and ¥0.020/kWh (1922) - 0.022/kWh (2000) for coal thermal power. As for the plants commissioned in 1992, nuclear and LNG thermal power generation can compete with each other when the exchange rate stands at ¥112/$.

Thus, compared with alternative generating sources, nuclear is influenced very little by exchange rate fluctuations. In the short run, it is not allowed to enjoy this advantage in the face of such fluctuations. On the other hand, however, this suggests that nuclear can provide a stable generating source in the long run against fluctuating exchange rates (Figure 5).

Summary

As we have seen so far, the relative economic advantages and disadvantages of different generating sources depend on fluctuations in such elements as the utilization factor, construction costs and exchange rate, with their sensitivity varying according to these fluctuations.

Among them, nuclear is shown to be very economical with high utilization levels. Nuclear running costs are also relatively insensitive to sharp exchange rate fluctuations. On the other hand, nuclear is more sensitive to rising construction costs than any of the other alternatives.

However, these construction costs are thought to decrease in the long run. If this were not the case and nuclear construction costs stay some 10% above the present level (at the plants commissioned in 1992), nuclear power would eventually become less economic than thermal power generation.

Figure 5: Generating costs by exchange rate

(When commissioned in 1992)

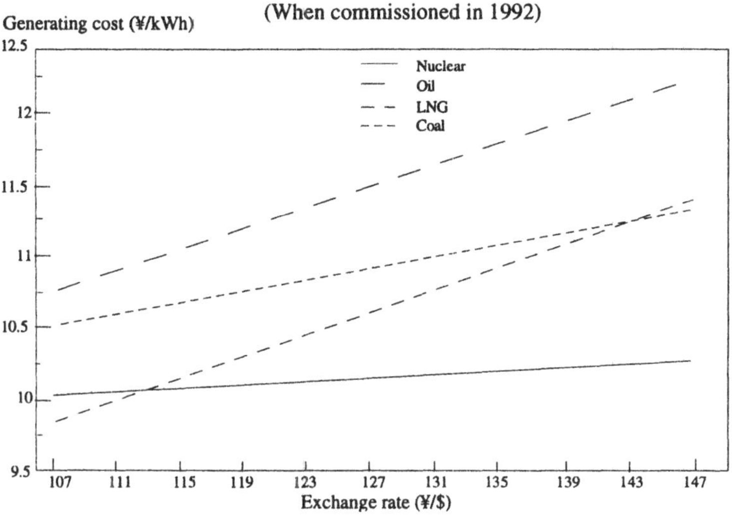

(When commissioned in 2000)

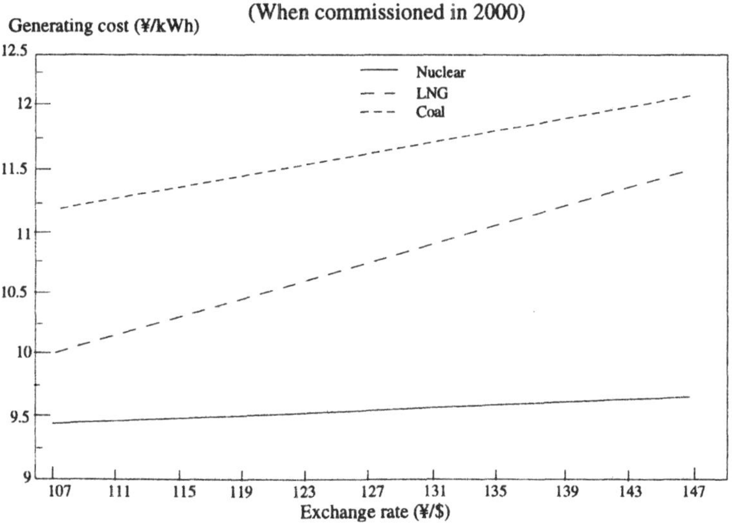

Cost of environmental control

In conventional generating cost estimation, cost of environmental control has been implicitly included in the cost of building equipment. Here, spending on SOx, NOx and CO_2 control are grouped together in environmental cost, and are made distinct from the generating cost incurred prior to any environmental control measures taken, in an attempt to show clearly the extra environmental cost.

In the case of nuclear power generation, the environmental cost required for SOx, Nox and CO_2 control is nil, and hence is far more advantageous in generating cost terms.

Environmental control cost is discussed here while referring to the estimations made by Messrs. Motofuji and Uchiyama of the Central Research Institute of Electric Power Industry.

Environmental control in thermal (coal, LNG-fired) power generation

For newly-built 1,000MW thermal power plants, one fired by coal, and the other by LNG, we consider three cases for control of flue gas from boilers. They are:

- No control, with flue gas discharged into the atmosphere as it is.
- SOx and NOx control. SOx and NOx control technologies assumed in making the estimations are coal-gypsum process (SOx) and ammonia catalyst contact reduction process (NOx); they are both currently in use in Japan.
- CO_2 control (two processes examined) on top of SOx and NOx control. As for CO_2 recovery, estimations are made for two technical options: amine process, of which commercialization is likely soon, and pure oxygen combustion process (PSA process), still being developed. In both cases recovered CO_2 in a liquid state is to be transported in specialized carriers to 3,000km offshore, then confined in 3,000m-deep waters.

Assumptions of construction and fuel costs, among others, are crucial to cost estimations. It is also necessary to specify the scope of environmental control.

Generating efficiency at newly built coal-fired 1,000MW thermal power plant goes down, from 38.84% in the no control case, and from 39.52% in the SOx/NOx control case, to 37.95%. With CO_2 control added, it goes down further to 25.88 - 29.14%. Reversely, generating cost goes up, from ¥8.11/kWh in the no control case, to ¥10.71/kWh in the SOx/NOx control case. With CO_2 control added, the cost reaches the ¥18.0-22.5/kWh range.

For the LNG-fired plant, generating efficiency changes little from 39.52% in the no control case to 39.44% in the NOx control case. But, with CO_2 control used, the efficiency deteriorates considerably to 31.68 - 32.62%. Generating cost amounts to ¥9.29/kWh in the no control case, which rises to ¥9.67/kWh in the NOx control case, and further to ¥15.0-15.9/kWh with CO_2 control added.

Figure 6: A menu of CO_2 stabilization measures

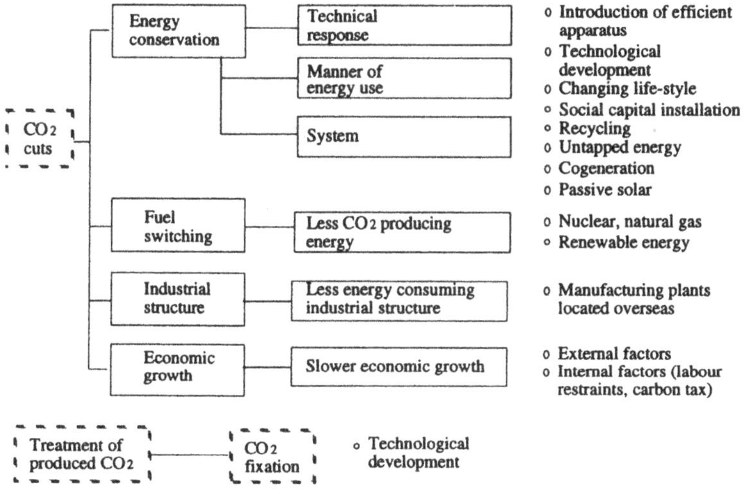

Tougher environmental standards, if introduced, would leave few choices but to incorporate the environmental cost into electricity rates. In this case, nuclear would become by far the most advantageous form of power. However, LNG thermal power could remain fully competitive within the scope of SOx/NOx control. But, if total CO_2 emission control standards (ex. CO_2/kW stabilization) were introduced, fossil fuel-fired thermal power could no longer rival nuclear. If it was allowed to incorporate the environmental cost into electricity rates, introduction of total CO_2 emission control would also become possible. But, given the nature of global warming, it is more realistic to consider how to check it in an ultra-long time span. Moreover, because the environmental cost incurred in total CO_2 emission control standards would cause a formidable burden to the public, such a short-range target as one by 2000 is nothing less than unrealistic.

Carbon tax to stabilize CO_2 emissions

Japan's Action Program to Arrest Global Warming (October 1990) set a target that per capita CO_2 emissions from 2000 onwards be stabilized generally at 1990 levels. What will be the size of a carbon tax, if it is introduced as the only instrument to attain the target? Figure 6 shows major options to stabilize per capital CO_2 emissions.

With such models as shown in Figure 7 in use, necessary carbon tax amounts are estimated.

Figure 7: Overall structure of energy econometric analysis model

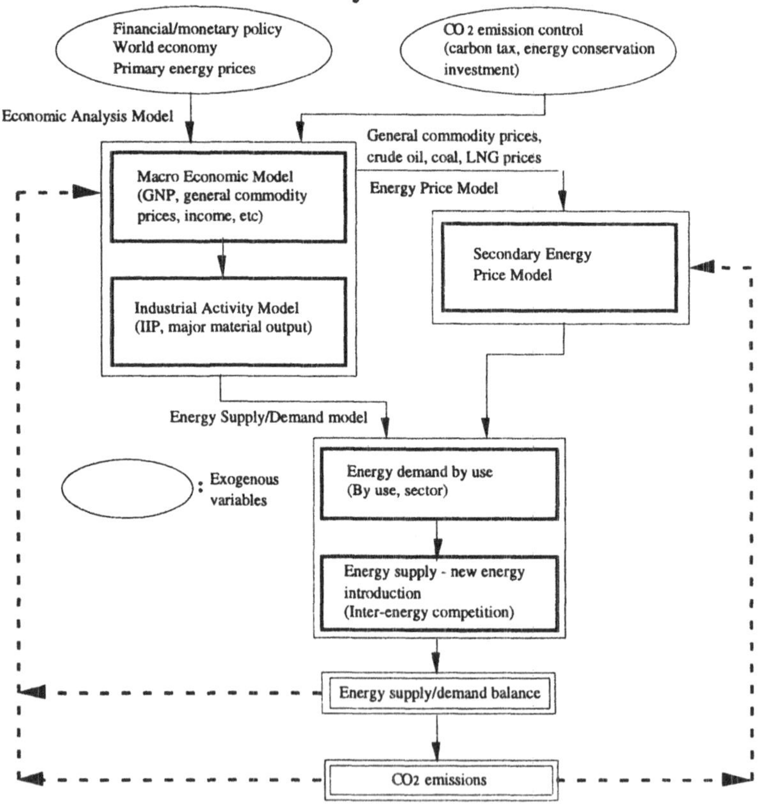

The carbon tax required to stabilize CO_2 emissions would amount to ¥46,200/kl or $70/bbl ($1=¥105) by 2000, and to ¥57,100/kl or $100/bbl ($1=¥90) by 2010. GNP growth would be down to 2.6%/year in 2000 from 3.7% in the base case, and down to 2.0%/year in 2010 from 2.6%/year in the base case. The growth of final energy consumption would slow down to 0.9%/year in 2000 from 2.7%/year in the base case, and to 0.4%/year in 2010 from 1.2%/year in the base case. This means GNP losses of as much as some ¥60 trillion ($56 million) in 2000, and of about ¥116 trillion ($1.2 trillion) in 2010. These suggest how huge the cost will be to stabilize CO_2 emissions. Moreover, it is nearly impossible to solve the issue in the short term. To eliminate this problem worldwide, including in developing countries, ultra-long-range efforts are needed by putting 2050 and afterward into perspective.

How to return carbon tax revenues, and results:

- The whole amount (100%) is to be returned, 25% each for lowering personal income tax, cutting corporate income tax, raising government spending (12.5% for government final consumption spending, 12.5% for public capital formation), and increasing subsidies.
- The rise in subsidies was considered in order to analyze strengthening of energy conservation investment later.
- Either of these ways of recycling, if used, results in a lower economic growth than in the base case.
- Depressed consumption by rising general prices can cause the real economy to shrink.
- Full amount of government tax receipts can hardly be recycled due to inevitable losses through individual channels.
- Given the purport of carbon taxation, it is hard to practise a recycling which produces particularly massive ripple effects.

Model-based simulation of carbon tax needed to stabilize CO_2 emissions

Objectives/aims of the simulation

- To examine size of carbon tax needed to stabilize CO_2 emissions.
- To examine economic impacts of carbon tax needed for CO_2 stabilization.
- To examine how individual functions of carbon tax work.
- To modify an 'Energy Econometric Analysis Model' developed by the IEE.
- and use it in the simulation.

- To back energy amounts, crude oil imports, etc., to macro economic terms and energy prices, then analyze interactions between the economy and energy.
- To make exogenous variables endogenous as much as possible.

How to impose carbon tax

- To start taxing from 1993 onwards by increasing the tax amount every year at a certain pace per carbon ton.
- In size, the carbon tax must be heavy enough to stabilize per capita CO_2 emissions in 2000 and afterward virtually at 1990 levels.
- Different energy sources are taxed by multiplying a per carbon ton tax amount by specific carbon emission coefficients to them, with resultant tax amounts added on their energy prices.
- Electricity and town gas are taxed by producing average carbon tax amounts by dividing total carbon tax amounts incurred in fossil fuel inputs, with resultant tax amounts added on electricity/town gas rates.

Discussion: Energy Pricing

Mike Parker

Most discussion of energy pricing (and other economic instruments) in the context of environmental protection is concerned with means, rather than ends. But this omits the vital question of determining appropriate objectives, striking the best balance between the costs and benefits of environmental protection. How can we use the price mechanism, whether by the use of taxes/charges or (negatively) by subsidies, as a means of determining the most appropriate environmental externalities for economic decisions?

In theory, the optimum level of a tax/charge (e.g. per unit of emitted pollutant) is that which minimises the sum of discounted damage and abatement costs over time, or, to put it another way, the 'correct' level of tax/charge (and amount of emissions) is that where the marginal value of reducing emissions equals the marginal cost of making that reduction (i.e. where the damage cost curves and the abatement cost curves intersect). This also means that a particular level of tax is only 'correct' if it is associated with the 'correct' level of emissions (or reduction in emissions).

But what are the difficulties in looking at the problem of setting targets in this way?

(i) There is no single set of damage/abatement cost curves: it depends on who you are and where you are. The equation will be affected by ability to pay, and therefore by income distribution both within the countries and (more significantly in the case of Global Warming) between countries. This is another way of saying that the trade-off between environmental benefit and economical cost will be quite different for the rich and for the poor.

(ii) There is no 'technical' way of deciding upon the appropriate discount rate to be used, which fundamentally affects the weighting of the long-term against the short-term. This is an issue of particular importance in determining investment. Price effects in the energy sector often have considerable time lags because of the long life of many energy producing/using assets and the lead times necessary to change the capital stock, which is a major route to environmental improvement.

(iii) There is great difficulty in drawing environmental damage cost curves because of scientific uncertainty, and problems of measurement. Moreover, the state of scientific knowledge is likely to keep changing. Even the shape (as distinct from the level) of these curves is in doubt - particularly for Global Warming.

(iv) There is also great difficulty in drawing abatement curves, since these are affected by uncertainty about the technological responses that may be available over time, and the extent to which 'abatement costs' can be mitigated by the effects of competition (i.e. whether a tax/charge could be passed on to the consumer or whether it represented a potential erosion of profit.) In addition 'abatement costs' in the widest sense must include opportunity costs of other economic benefits (i.e. reduction in activities which would otherwise have been profitable).

(v) It is not the tax/charge itself which produces the desired market response, but the <u>post-tax price</u>: that is the tax in combination with the pre-tax 'commercial' price. The 'commercial' price is therefore an essential part of the overall formula. But this in turn raises a number of questions (particularly in the case of fossil fuels):

   (a) 'Commercial' prices vary considerably over time, depending primarily on the balance between demand and capacity (SRMC and LRMC are not pricing rules in competitive markets, but the <u>result</u> of supply and demand balances). The addition of a fixed environmental tax would therefore lead to a fluctuating post-tax price.

   (b) Existing prices are subject to major distortions introduced for non-environmental reasons - e.g. OPEC, German coal subsidies, taxation of North Sea economic rent. This raises the question whether or not other tax/subsidies should be subsumed within environmental taxes (Nick Hartley has referred to the high level of taxes already levied on oil).

   (c) 'Commercial' prices are increasingly international, so that environmental taxes which were (correctly) differentiated between countries would lead to differentiated post-tax prices, which raises the whole question of international competitiveness.

(d) Other things being equal, environmental taxes will reduce commercial' prices whether absolutely (if applied to all energy) or relatively (if differentiated between fuels). This in turn reduces the impact of a particular level of tax, which would then have to be ratcheted up to compensate for the feedback effect. But this would be a very difficult process to manage.

The above considerations suggest strongly that no universal or permanent 'correct' values for environmental taxes or emission limits can be arrived at, and that the practical problems of arriving at even 'second-best' approximations are very great. But, it might be argued, this means that nothing would be done, even where it is obvious that something should be done. Clearly, where the benefits of a course of action are large and obvious, and where the costs are small and readily acceptable (as in the case of most issues of public safety) then there is no point in agonising about the theoretical optimum solution. However, many cases are not clear cut, and given the uncertainties and difficulties outlined above, decisions by the authorities on 'tax' values and acceptable emission levels will be largely 'political'. Both George Yarrow and Nick Hartley have referred to the political nature of many of these decisions: particularly where redistribution is involved. But should politicians err on the side of maximising the effectiveness of emission controls, or on the side of minimising economic impact? And how will <u>that</u> be decided, other than by the relative volumes of noise from interested parties? In other words, is rationality possible even with 'second-best' solutions, or do governments have little but politics to guide them? The fact that government action may be stated in terms of the precautionary principal does not get rid of the difficulty - how should the correct level of the insurance premium be decided? Is the best possible solution that which is the least unpopular?

George Yarrow has pointed to the particular difficulties of devising appropriate policies to deal with global warming. It seems to me that the difficulties outlined above make it virtually impossible to arrive at economically optimal 'taxes' and emission limits for greenhouse gases. I have much sympathy with the philosophical difficulties on global questions set out earlier by Dr Eyre. With global warming, conventional economics and the price mechanism seem to be of little help in deciding the targets; and <u>in this case at least</u>. I doubt whether Nick Hartley is right in concluding that 'ultimately environmental problems are economic problems which are susceptible to classic economic remedies'.

So where does this bring us? I conclude that the conventional view of the optimal price mechanism, whereby marginal price equals marginal cost, is of little value in discovering the optimal level of energy taxes or emission limits - least of all for global warming. Of course, this does <u>not</u> mean that the price mechanism cannot be an effective (and most often the most effective) way of meeting a particular environmental target. In particular, the case for tradable permits is often a strong one. But this requires somebody to make the prior judgement either on the target of allowable emissions, or on the 'tax' that will be applied. On <u>target setting</u> we still have to find ways of bringing rationality to a second-best political process.

PART 4

Public Policy Towards New Investment

CHAPTER 9

Regulatory interventions for promoting investments in environmentally benign energy technologies

Kenji Yamaji

Introduction

There are different views on the role of governmental interventions as a policy tool. Economists tend to favour market based measures believing in the invisible hand of the market to achieve optimal allocations of resources. They try to minimise governmental interventions ('command and control') which distort market functions. Practitioners, on the other hand, advocate governmental interventions as indispensable measures to correct market imperfections. In their views, governmental interventions are not so rigid as implied by 'command and control', but are a soft package of practical policy tools.

In the field of energy and environmental regulation, particularly those measures proposed for coping with climate change, heated discussions are now being conducted on the effectiveness of various policy options. Policy measures to cope with climate change vary widely. They include both market based tools such as carbon tax, and many governmental interventions for promoting specific actions such as energy conservation.

This paper explores the role of governmental interventions for promoting investments in environmentally benign energy technologies through examining cases in Japan. In the following, measures taken in Japan to promote investments for efficient energy use are reviewed; then measures for promoting investments in non-fossil energy sources are explained; and, in the last part, a study on economic measures for CO_2 reduction is presented.

Figure 1
Changes in energy consumption per unit GNP

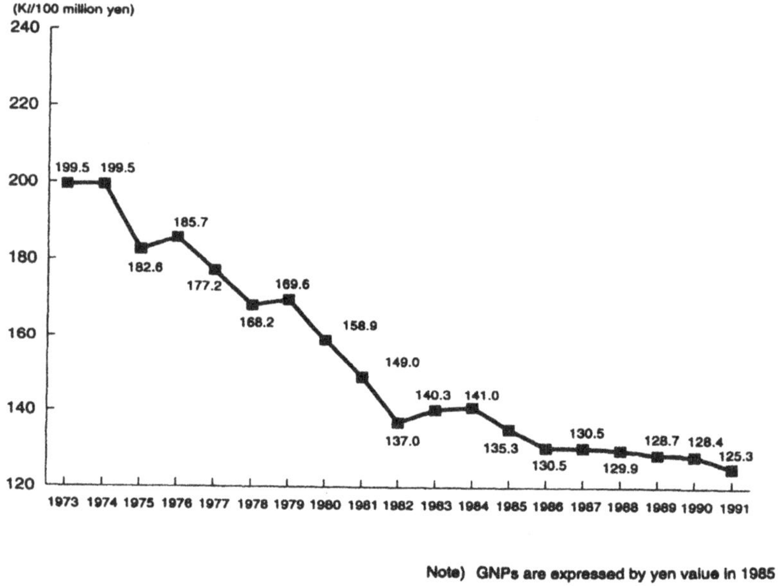

Source: The Energy Conservation Center, Japan Energy Conservation Handbook 1992

Measures to promote investments for efficient energy use

The basis of the policy for energy conservation in Japan was set by the Law Concerning the Rational Use of Energy, in 1979. According to the law, standards for factory energy management, housing insulation, and automobile fuel efficiency are stipulated. Education and information dissemination on energy conservation are also promoted systematically along with the promotion of technology development related to energy conservation. As a result, as shown in Figure 1, energy consumption per unit in Japan has improved by approximately 37% since the first oil crisis in 1973.

Figure 2
Investments for energy conservation in Japan

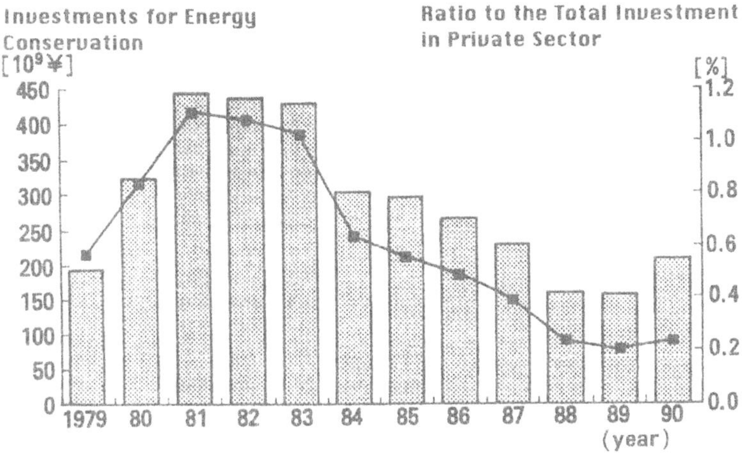

The investment for energy conservation was stimulated, as shown in Figure 2, by these policy measures together with high energy prices in the early 1980's; however, the effect faded in the latter half of the decade. Then, the climate change issue brought about a new impetus to encourage energy conservation. The government of Japan enacted two laws in 1993 to reinforce the Law Concerning the Rational Use of Energy: the Energy Supply and Demand Structure Advancement Law, and the Energy Conservation and Recycling Support Law. The targets for energy conservation were raised and extended to many areas through these amendments.

The government encourages and guides business sectors to follow the reinforced regulations. Those who fail to meet the lowest targets would be penalised while those clearing the new, stringent targets would be given tax incentives and/or other financial benefits shown in Table 1. The Ministry of International Trade and Industry (MITI) requires some 3200 large factories, which represent a dominant part of the total energy consumption by the industrial sector, to report their fuel and electricity consumption. MITI has jurisdiction to order these designated factories to keep conservation targets. The ministry would be able to publish the names of firms which violate these requirements.

Table 1: Legislation for energy conservation in Japan

	Measures to reinforce the energy conservation effort*	Measures to support the energy conservation effort**
Factories	(a) Energy conservation targets (set by government) (b) measures to ensure the compliance with the government set targets by designated factories - Duty of periodic reporting to MITI - Guidance and command by MITI to factories which cannot achieve the targets - Penalties (publish the names of non-compliant factories, impose a fine)	Financial assistance to designated enterprises - Low interest loan and interest rate subsidy program (4.05%) - Debt guarantee by a special fund - Special tax system for promoting energy conservation investment (7% tax-credit, or 30% accelerated depreciation of assets)
Buildings	(a) Energy conservation targets (set by government) (b) Equipment for which the rational use of energy should be promoted (air-conditioning, hot water supply, lighting, etc) (c) Measures to ensure the compliance with the government set targets by designated constructors	Financial assistance to designated enterprises - Low interest loan and interest rate subsidy program (4.25%) - Debt guarantee by a special fund
Machines and Appliances (including automobiles)	(a) Energy conservation targets (set by government) (b) Labelling of energy consumption efficiency	-
Technology Development	-	(a) Subsidies to the development of energy conservation technology (b) Financial assistance to designated enterprises - Low interest loan and interest rate subsidy program (4.1%) - Special tax system for promoting energy conservation investment (6% tax-credit)
Area-wide Energy Conservation	-	Area-wide energy conservation type supply system - Low interest loans by the Japan Development Bank - Debt guarantee by a special fund

* Measures based on Energy Supply and Demand Structure Advancement Law
** Measures based on Energy Conservation and Recycling Support Law

Source: H Imura, private communication, 1993

Figure 3
Improvements in energy efficiency of home electric appliances

(1) Refrigerators

Indices of electric power consumption for one month (annual average) of the 2-door, 170-liter type
(1973=100)

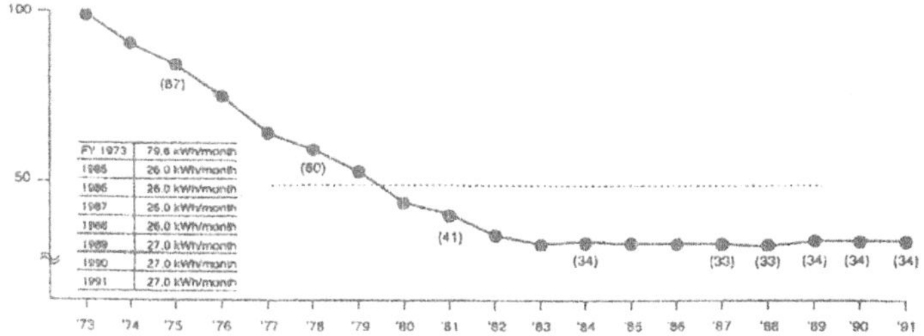

(2) Color television sets

Indices of electric power consumption of the 19, 20-inch type
(1973=100)

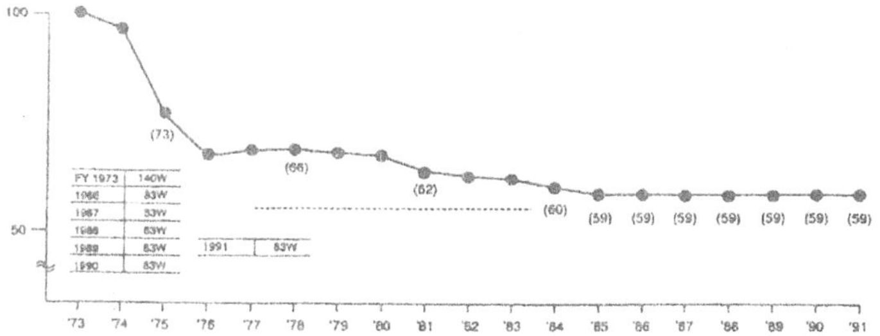

(3) Room air conditioners

Indices of electric power consumption of the separate type
(1973=100)

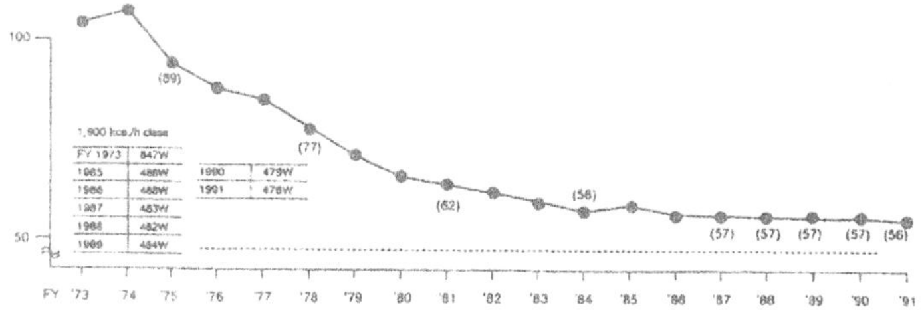

Source: The Energy Conservation Center, Japan Energy Conservation Handbook 1992

Improvements in energy efficiency of home electric appliances were, as shown in Figure 3, mostly achieved by the early 1980's. This trend is also found in fuel efficiencies of automobiles. While the government is preparing new extended conservation targets, we should recognise that energy efficient improvement in household and transport sectors is a challenging problem.

Effectiveness of government interventions critically depends on administrative capabilities. For the setting the targets or standards which are effective and feasible, administrative authorities must have detailed knowledge about individual technical conditions in each sector regulated. The authorities must also have infrastructures for monitoring and inspection of compliance with the standards or targets. In this regard, relations between the government and business sectors in Japan provide a favourable condition for the enforcement of effective and efficient interventions. The human network connecting the government and private sectors in Japan has so far formed in a producer oriented way. However, the network has an inherent weakness in intervening in consumers' behaviour.

Measures to promote investments in non-fossil energy sources

Figure 4 shows the historical trend and projection of component shares of primary energy supply in Japan. Since the first oil crisis, the share of oil has reduced from 77% to 57%, and the reduction was replaced almost evenly by nuclear power and national gas. The share of nuclear power is expected to increase further in future while that of natural gas is assumed to increase modestly. Solar and other new renewable energies, which are included in 'Others' in Figure 4, are expected to increase their shares, but their absolute contribution will remain very small even in 2010.

Nuclear power has been chosen and intensely promoted in Japan as the primary energy to replace oil and to meet future demand growth. While the technological performance of nuclear power in Japan is very good and its economic competitiveness in electricity generation is kept firmly in step with the conditions of low fossil fuel prices, the siting of new nuclear power stations poses a potential bottle-neck for further development.

To promote power plant siting, particularly for nuclear power plants, a significant amount of government budget is used. As shown in Figure 5, which describes money flows related to energy taxes in Japan, the revenue collected through the Electric Power Development Promotion Tax which is levied on electricity sales, is supplied to the Special Account for Electric

Power Development. About half of the special account is used for the promotion of power plant siting such as special grants for regional government where power plants are located. The budget amounted to around 190 billion yen in 1992.

Figure 4

Changes in primary energy supply component shares

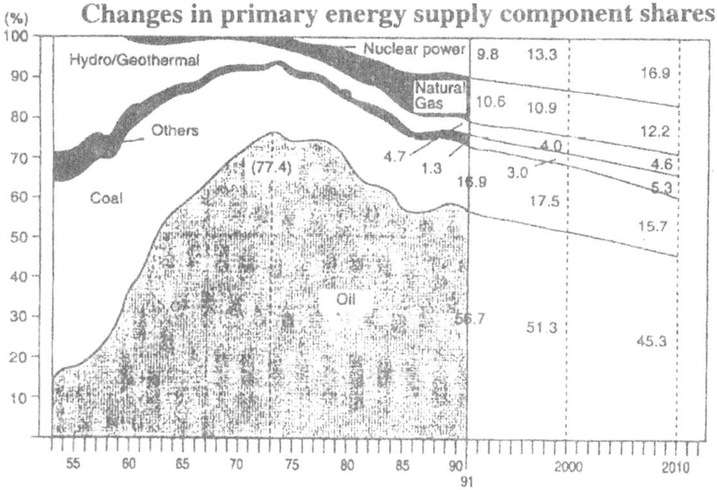

Note) The values for the 2000 and 2010 fiscal years were based on the energy supply and demand projection which was the basis for the oil substitute energy goal (determined at the cabinet meeting on October 30, 1990).

Source: The Energy Conservation Center, Japan Energy Conservation Handbook 1992

It is also shown in Figure 5 that a significant portion of energy R&D budget is supplied by the special account. Out of the total energy R&D budget of around 400 billion yen in 1992, only 159 billion is supplied by general account and the remaining portion is supplied mainly through the Power Source Diversification Account in the special account. A dominant part of the energy R&D budget is used for the development of advanced nuclear technologies such as Fast Breeder Reactors.

There are also policy measures available to promote renewable energies. Some portion of the above mentioned special account is used for R&D into renewable energies, and subsidies to promote market concentration of decentralised energy sources such as photo-voltaic cells. Some local governments also have similar arrangements to waive tax or to compensate some portion of investment costs for promoting renewable energies. In addition, electric utilities are guided to purchase surplus power generated by photo-voltaic cells and windmills at the retail prices of electricity. These measures will lower the initial barriers to commercialisation and lead to cost reductions for renewable energy sources.

Figure 5
Flow of energy related tax revenues in Japan

(unit: 10^9¥ in FY 1992)

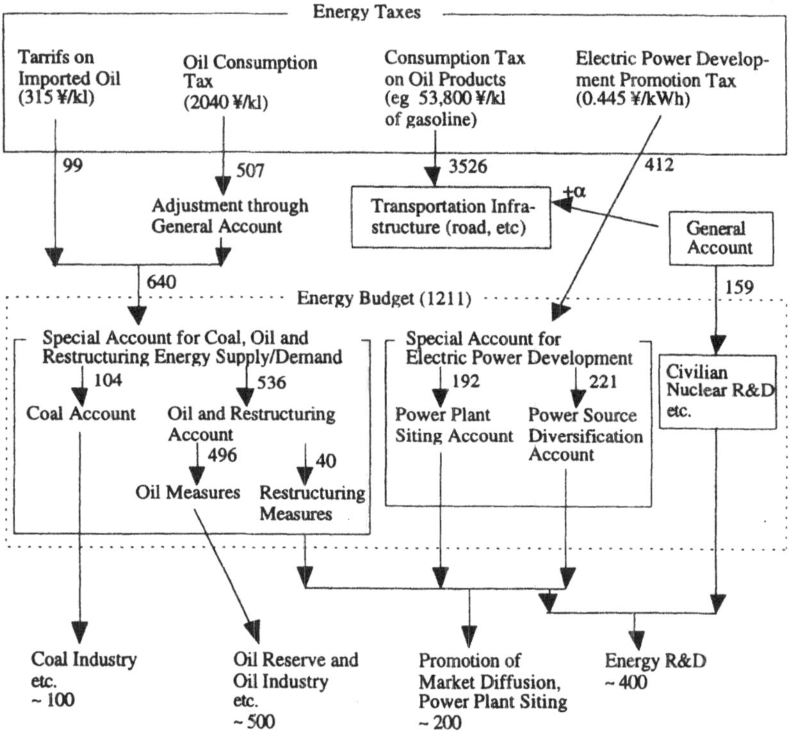

Assessment of economic measures for CO_2 reduction in Japan

An outline of a study on a carbon tax and a subsidy for CO_2 reduction is presented in the following. (For details, refer to K. Yamagi et al. 'A study on economic measures for CO_2 reduction in Japan'. *Energy Policy*, 21, 2, pp. 123-132, 1993).

Figure 6
CO_2/Energy/GNP Growth Rate Indicators in Japan

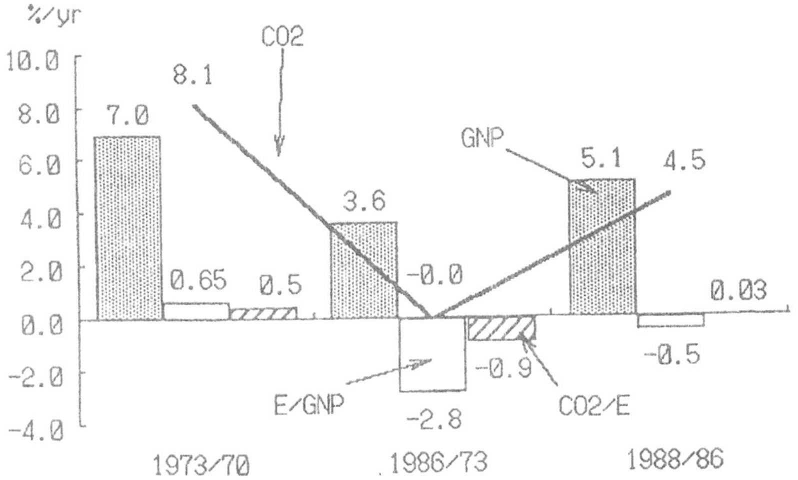

Stabilisation of CO_2 emissions in Japan

Figure 6 shows the CO_2/energy/GNP growth rate indicators for the period 1970-1988 by dividing it into three sub-periods; 1970-73, 1973-86, and 1986-88. CO_2 emissions in Japan were kept constant during the sub-period of high oil prices (1973-86); however, CO_2 emissions had increased more rapidly than the economic growth rate before the first oil crisis and have been again increasing almost in parallel with GNP since 1986. Thus, it would be too optimistic to extrapolate the performance shown in the period of oil crises to the future.

Figure 7 shows the base case projection by energy related CO_2 emissions to the year 2005 calculated by the Central Research Institute of Electric Power Industry (CRIEPI). In this projection, primary energy shares of nuclear power, natural gas, coal and oil in 2005 are 15%, 13%, and 47% respectively; and, CO_2 emission in 2005 increases by about 36% from the level of 1988. This base case projection is used as a reference for simulation studies to reduce CO_2 emissions.

Figure 7
Base case projection of CO_2 emissions in Japan

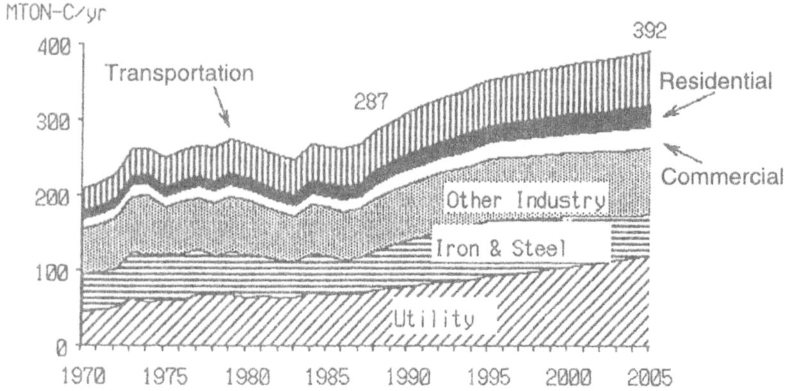

Methods for assessing policy measures to reduce CO_2 emission

There are generally speaking two types of approaches for assessing the effects and costs of measures to reduce CO_2 emissions. One is a model-based approach which employs macro-economic models as a key role to evaluate the total impacts of the measures on energy systems and national economy. The other is a so-called bottom-up approach which identifies individual technical measures and accounts individual costs and effects. For the assessment of subsidy policy we apply an integrated system combining the two approaches while only the part of model-based approach is used for assessing CO_2 tax policy.

Analysis of CO_2 tax policies

According to the CRIEPI's base case projection of CO_2/energy/GNP, average growth rates of GNP, energy, and CO_2 emission for 1988-2005 are 3.7%/year, 2.1%/year, and 1.8%/year respectively. Starting from this base case projection, let us assume the introduction of a CO_2 tax to hold CO_2 emission in 2005 to its 1988 level. The CO_2 tax is to be introduced in 1990, and the level of tax is increased gradually to maintain the effect of the tax.

We assume that the CO_2 tax is imposed on primary energy sources in proportion to theoretically estimated CO_2 emissions. Both the direct impact of CO_2 tax through raised fossil fuel prices and indirect impact through changes in economic outputs and general prices are evaluated by the CRIEPI's Medium-Term Economic Forecasting System. Depending on the ways of using the revenue of the CO_2 tax, we present two cases: Case 1 'Tax Remove Case' and Case 2 'Tax Offset Case'. In Case 1 collected tax revenues are assumed to be removed from the Japanese economy; and, in the Case 2 the revenue of the tax is offset by the equivalent amount of income tax reduction.

Figure 8
Carbon tax for holding CO_2 emission constant

- Introduce in 1990 with a level of ¥4,000/t-C
- Then, the tax level increases yearly by ¥4,000 afterwards, i.e. ¥64,000/t-C in 2005

e.g.

Crude Oil Price (Nominal)

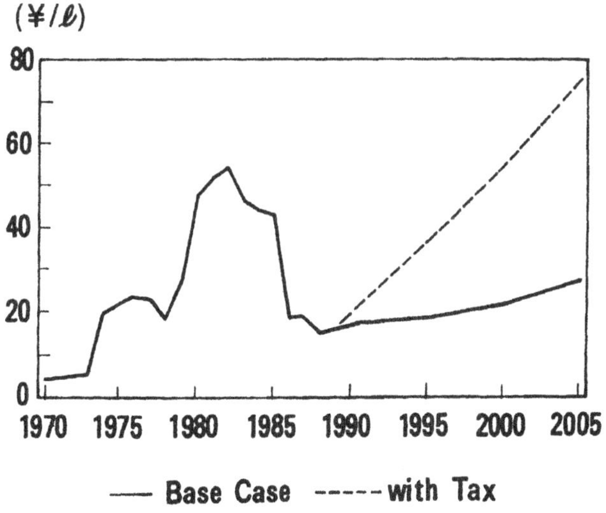

Through several simulation experiments we found the tax schedule to hold the CO_2 emission in 2005 approximately at the level of 1988. As stated in Figure 8, the tax schedule to achieve the target is as follows: CO_2 tax of 4,000 yen per ton of carbon is introduced in 1990, and then increased yearly by 4,000 yen until it reaches 64,000 yen per ton in 2005. As also shown in Figure 8, the CO_2 tax raises oil price in 2005 to the level of about 2.8 times as much as that in the base case projection. Reductions of CO_2 emissions caused by the tax are illustrated in Figure 9.

Figure 9
CO_2 Emissions in 2005

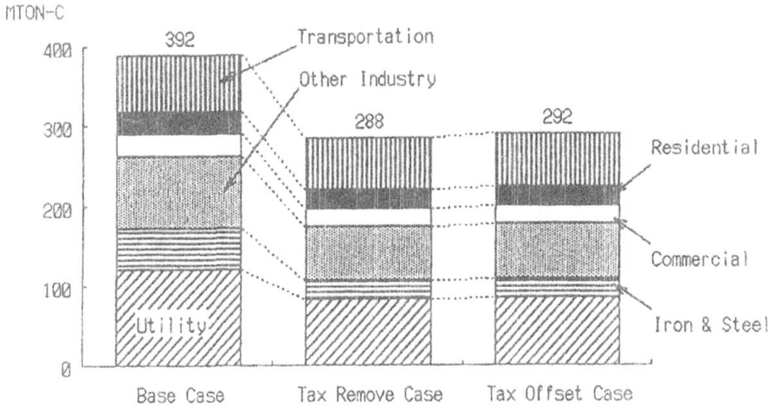

The macroeconomic impacts of CO_2 tax policy to hold the CO_2 emission constant are summarised in Table 2. The results lead to the following findings:

- CO_2 emission stabilisation through CO_2 tax has a serious impact on national economy: in the Tax Remove case, average growth rate of GNP in real term from 1988 to 2005 reduces by 0.4%/year and GNP in 2005 decreases by 38 trillion yen (1980), more than 6% from the base case value.
- Income tax reduction offsetting the CO_2 tax revenue mitigates the adverse macroeconomic impacts: loss of GNP in 2005 in the Tax Offset case is reduced by 30 trillion yen (1980), about 5% of the base

case value; however the effect is too small to make CO_2 tax an attractive policy to Japan.

- GNP losses per unit CO_2 reduction range from 250 thousand yen/ton-C (cumulative average in the Tax Offset case) to 360 thousand yen/ton-C (2005 year only in Tax Remove case). The specific CO_2 reduction costs associated with CO_2 Tax in Japan are prohibitively high when compared with the cost estimates of various technical measures to reduce CO_2 emissions, mostly less than 50 thousand yen/ton-C.

Table 2: Cost and effects of CO_2 Tax

	Remove	Offset
A. Real GNP Loss (Trillion ¥)		
in 2005	-37.7 (-6.2%) *2	-30.0 (-4.9%)
1990~2005. Cumulative	-310.6 (-4.0%) (-0.4%, Annually)	-230.6 (-3.0%) (-0.3%, Annually)
B. CO_2 Reduction (Mil. C-ton)		
in 2005	104.1 (26.6%)	99.8 (25.5%)
1990~2005. Cumulative	985.1 (17.1%)	939.0 (16.3%)
C. Tax Revenue (Trillion ¥)*1		
in 2005	11.9	12.0
1990~2005. Cumulative	118.2	119.0
D. Specific CO_2 Tax Revenue (Thousand ¥/C-ton)		
in 2005	114	120
1990~2005. Cumulative	120	127
E. National Economic Cost per CO_2 Reduction (Thousand ¥/C-ton)		
in 2005	362	301
1990~2005. Cumulative	315	246

*1 Deflated by wholesale price index (Real Price in 1980)

*2 Difference rate from Base Case

Analysis of Subsidies for Reducing CO_2 Emissions

An integrated system combining a bottom up approach with CRIEPI 's Medium -Term Economic Forecasting System is used for assessing the effect of the subsidy. First, we discuss the economic criteria for employing measures to reduce CO_2 emissions. Then we evaluate CO_2 reductions by the subsidies, and macroeconomic impacts of the subsidies.

We need to introduce some economic incentives to accelerate the implementation of various technical measures for CO_2 reduction. Here we assume the subsidy Pc for CO_2 reduction as this economic incentive. We express Pc in terms of yen per ton-C reduced. With this subsidy Pc, the fundamental economic condition for employing specific technical measures is stated as follows:[savings from reduced energy consumption] + Pc* [ton C reduced] is larger than the total cost of implementation. In the case of investment decisions, this condition leads to so called pay back time criteria, namely, the ratio of investment to expected annual benefit. Since energy prices have fluctuated widely in the past, entrepreneurs tend to be rather conservative when they invest in energy equipment. In Japan, they typically use a short payback time of around 2 years. We used this criteria in the simulation.

Before investigating individual measures, we should differentiate between two types of energy conservation measures. One type introduces new facilities so as to reduce energy loss in processes. For example, cokes dry quenching systems used in iron and steel industries belong to this type. For convenience, we call this type the investment type. The other type changes the usage patterns of machines or materials in order to introduce economies in the system. Typical examples of this type are changes in the usage patterns of automobiles, i.e. running distances which is caused by changes in the fuel prices of automobiles. We call this type the pattern change type. We assume that measures of the pattern change type are not implemented by introducing the subsidy. On the other hand, the investment type measures can be given economic incentives to be implemented in the market by the subsidy. The investment type measures are difficult to evaluate without a detailed survey on the characteristics of such measures, which is the essence of bottom-up approaches.

The first step in evaluating measures is to determine the amount of subsidy Pc so that these measures can be implemented by actual market incentives. Given the cost data, the Pc necessary for each measure to be marketable can be determined. Here we utilise data gathered mainly by the Japan Energy Conservation Center, which is not a thorough survey of the results but only the data presently available. Measures in the data include 57 energy conservation technologies in industries, thermal insulation for houses and buildings, and fuel switchings in industrial boilers/ furnaces and in electric power stations. For an illustrative example, engineering-economic characteristics of industrial energy conservation technologies are shown in Figure 10.

Figure 10
Supply curve for energy conservation technologies

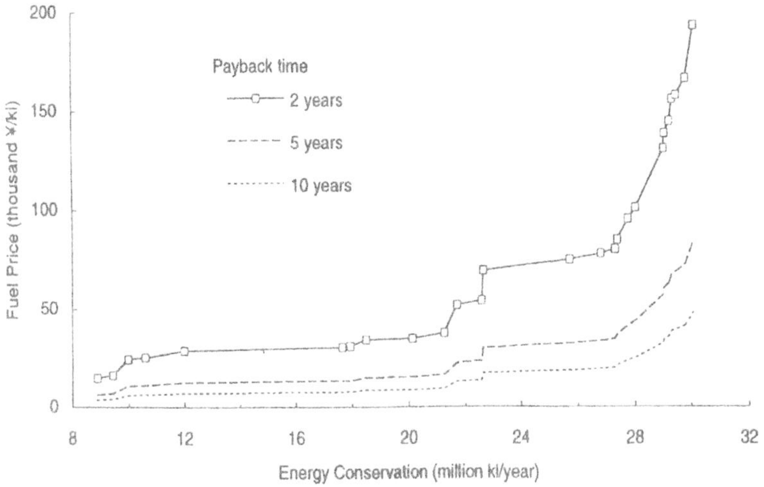

Even if the marketability of measures is guaranteed, it takes time for that measure to be effective due to the time needed for replacing old facilities. Taking this into account we made the following assumptions about the speed of acceptance for marketable measures :

- The market penetration rate for energy conservation measures is assumed to be 10%/year. In other words an old facility will be fully replaced by a new one in 10 years.
- Taking this into account the long lifetime of houses and buildings, we assume that the market penetration rate of insulation materials for houses is 40 years.
- As for electric power plants, we decide the time needed for replacing each plant based on its vintage assuming that the average life time of a plant is 35 years.

Based on these assumptions, we evaluate how much reduction in CO_2 emissions will be attained by the introduction of a Pc. Here we should be aware of the fact that only measures of the investment type are considered in the analysis. The results are shown in Figure 11 and Table 3.

Figure 11
Sectoral CO_2 reduction by subsidy in 2005

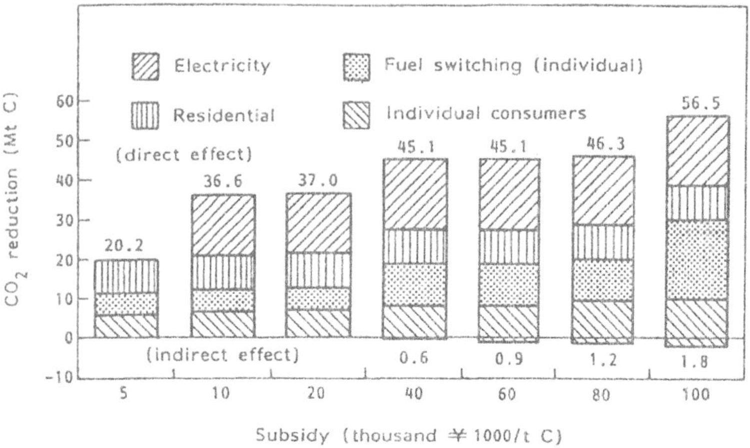

Table 3: Cost and effects of subsidy for CO_2 reduction

Level of subsidy (10^3 ¥/C-ton)	5	20	80
A. Real GNP Gain (Trillion ¥)			
in 2005	0.05 (0.01%)	0.33 (0.05%)	1.66 (0.27%)
1989~2005. Cumulative	0.50 (0.01%)	3.50 (0.04%)	22.65 (0.28%)
B. CO_2 Reduction (Mil. C-ton)			
in 2005	20.2 (5.1%)	37.0 (9.4%)	46.3 (11.8%)
1989~2005. Cumulative	150.9 (2.5%)	267.3 (4.4%)	426.4 (7.0%)
C. CO_2 Increase by Subsidy (Mil. C-ton)			
in 2005	0.03 (0.01%)	0.24 (0.06%)	1.20 (0.31%)
1989~2005. Cumulative	0.33 (0.01%)	2.32 (0.04%)	14.96 (0.24%)
D. Subsidy (Trillion ¥)*			
in 2005	0.09	0.62	3.11
1989~2005. Cumulative	0.71	5.02	32.42
E. Unit CO_2 Reduction Cost (Thousand ¥/C-ton)			
in 2005	4.2	16.9	68.9
1989~2005. Cumulative	4.7	18.9	78.8

* Deflated by wholesale price index (Real Price in 1980)

As shown in Figure 11, the marginal effect of the subsidy on CO_2 reduction becomes small as the subsidy rate Pc increases and indirect effects of subsidy, namely positive impact on macroeconomies and associated CO_2 increase is relatively small. But, please keep in mind that financing of the subsidy is not considered in this simulation. When the fund for the subsidy is raised through taxation, there would be a negative impact on the macroeconomies. Table 3 shows that the fund required for the subsidy Pc=20 thousand yen is 0.62 trillion yen in 2005; and 37 million ton-C is reduced with this subsidy. Thus, the required money per ton-C reduction is around 17 thousand yen. Compared with the results of carbon tax in Table 2, it is clearly shown that subsidy is a more efficient policy tool than carbon taxes to reduce CO_2 emissions in Japan.

Observations

The findings in the study on the economic measures to reduce CO_2 emissions in Japan are summarised as follows:

- Without CO_2 limitation policies, CO_2 emission from Japan's energy system would increase significantly : 36% during 1988-2005 according to CRIEPI's Medium-Term Economic Forecasting System.
- The level of CO_2 tax rate to stabilise Japan's CO_2 emission only through price effects is very high: primary fossil fuel prices should be 2-5 times higher than those in the base case in 2005.
- The national economic impact of CO_2 emission reduction through CO_2 tax is significant: GNP in 2005 decreases more than 6% for holding the CO_2 emission to the level of 1988 when the revenue of the CO_2 tax is removed from the Japanese economy.
- The national cost of CO_2 reduction by the price signal of carbon is prohibitively high: GNP loss per unit CO_2 reduction is around 300 thousand yen/ton-C.
- Income-tax reduction offsetting CO_2 tax revenue mitigates the above adverse impacts; however, the effect is too small to make CO_2 tax an attractive policy to Japan.
- The level of subsidy required to reduce about 40 million ton-C of CO_2 emission in 2005 is relatively small: around 20 thousand yen per ton of carbon reduction. But the marginal effect of the subsidy decreases as the level of subsidy increases, thus CO_2 emission in Japan cannot be stabilised by introducing the subsidy only.

- The national economy is stimulated by introducing the subsidy. But the associated increase in CO_2 emission is small: less than 0.1% for the subsidy level of thousand yen/ton-C.

There still remain many issues to be analysed. Among them, we identify the following points:

- To evaluate the effect of using the tax revenue as a subsidy to accelerate market penetration of specific technical measures for reducing CO_2 emission. In the study analyses of CO_2 tax and subsidy were done separately and the financing of the subsidy was not taken into account.
- To investigate the influence on global CO_2 emission through international economic impacts caused by the structural change of the Japanese economy to reduce Japan's CO_2 emission.
- To clarify what happens if many countries introduce CO_2 tax simultaneously.
- To identify the advantage which could be achieved through regional and/or global interactions to reduce the global CO_2 emissions more efficiently, e.g. technology transfer and international trade of v emission rights.

Through the above simulation study, it is demonstrated that governmental interventions such as subsidies to promote individual CO_2 reducing investments could be a more efficient policy tool than simple straight economic measures such as a carbon tax. We should be, however, very cautious in actual policy implementations bearing in mind that the effectiveness of governmental interventions critically depends on the administrative capabilities .

CHAPTER 10

Environmental regulation, investment and technical change

Jim Skea

Introduction

The regulation of economic activity which has adverse environmental impacts is shaped by a wide range of factors, including the extent of damage caused to humans and natural eco-systems, the technology available to prevent or control pollution, and perceptions of the wider economic and social consequences of mitigation actions. While the importance of a broad perspective is acknowledged, this chapter focuses specifically on the way in which environmental regulation may stimulate or inhibit the development and diffusion of cleaner production and transformation processes in the energy sector. It addresses also the closely linked question of the economic efficiency of regulatory instruments.

The paper uses two conceptual building blocks. The first is a vintage 'model' of investment and stock replacement which is used to conceptualise decisions about new plant investment, retrofitting and retirements.[1] The second is the extensive literature on the selection of regulatory instruments, e.g. standards versus taxes.[2] Taken together, these two basic elements provide a powerful framework for examining the interaction between investment and technical change on the one hand, and the selection and design of regulatory instruments, the stringency of controls and the influence of the competitive nature of the regulated market on the other.

The energy sector is associated with many environmental problems arising from production, transformation and consumption. It is therefore difficult to generalise about all the forms of regulation to which the sector is subject. While the paper is intended to be as general as possible, it focuses on production/transformation processes with the specific example of the regulation of conventional combustion emissions, sulphur dioxide (SO_2) and nitrogen oxides (NOx), in mind. The analysis illuminates a range of issues taken into account during regulatory processes, including: the capacity of regulated firms to raise investment capital; the broader economic consequences if significant capital expenditure is entailed; the indigenous capacity within a country to supply compliance technology; and the degree to which regulations may stimulate technological change, giving competitive advantage either to individual suppliers or to countries as a whole.

The paper begins with a brief description of the vintage 'model' of investment. The principal features of the various regulatory instruments which have been (or might be) used to influence combustion emissions in the energy sector are then described drawing on specific examples drawn from experience in the US, the UK and other parts of Europe. The following section considers more briefly the importance of the institutional framework within which regulations are developed and the way in which market organisation and structure may influence both the development of regulations and the consequent incentives for investment. The final section considers in more detail some specific issues and problems arising at the regulation/investment interface. Questions covered include: the rate of replacement of plant; problems of lock-in to old technologies; the degree to which regulations are robust with respect to technological and market change; and asymmetries of information between regulators and the regulated. Some emphasis is given to negative lessons ('how not to do it') and the occurrence of regulatory perversities.

The framework for investment

At any given time, a market (for electricity for example) will be supplied by plants of different vintages embodying technologies with a range of cost and environmental characteristics. As a result of technological change, the characteristics of potential new plants may differ from those of the stock of existing plant. Incremental technical change may have improved the performance of existing plant types, while more radical technological change may have led to the development of new types of plant with very different characteristics.

Investment in new plant will be induced by: a) the expectation of expanding markets; or b) the expectation that costs can be reduced by replacing older vintages. With energy demand growing rather slowly in OECD countries, the importance of replacement investment has grown in relation to investment aimed at meeting expanding markets. For replacement investment, the key issue is whether the avoidable costs of continuing to operate old plant (excluding unavoidable items such as interest charges) exceed the total costs of building and operating new plant. The balance between the costs of old and new plant can be altered by: the physical deterioration of existing equipment; changes in markets for fuels and raw materials; technological change which enhances the attraction of new investment; and regulatory changes (environmental or otherwise) which have differential effects on different plant types.

At the technical level, the response to environmental regulation may be divided into the following categories:

- replacing existing plant with alternative production techniques;
- ceasing production;
- investing in research, development and demonstration (RD&D) activity to make available techniques with better environmental performance at a later date;
- changing raw material and/or energy inputs to existing plant; or
- modifying existing plant by retrofitting 'end-of-pipe' control equipment.

Whether the modification of existing plant is appropriate or not depends on the balance of costs between that option and investment in new plant. However, in general, retrofitting an existing plant is almost always more expensive than the incremental cost of incorporating similar clean-up at a new plant built from scratch. Equally, changing raw materials or fuel inputs at an existing plant generally involves capital expenditure which would not be entailed had the same raw materials or fuel been used from the beginning. From the point of view of costs/investment, embodying higher standards of environmental performance in line with the 'natural' rhythm of plant replacement is an attractive option. However, whether this is ultimately desirable depends on the wider question of the balance between abatement and environmental damage costs. There is often a mismatch between the natural life-cycle of plant replacement, and the demand for much shorter response times set by the political process.

Across an entire market, a mixture of different technical responses is likely to be adopted in response to a specific regulatory measure. The precise nature of the response will depend on: the cost and availability of various technologies; the existing capital stock of plant, its vintage and its performance in terms of cost and environmental impact; the nature and structure of the regulated market (e.g. competitive versus regulated monopoly); and the technological capacities of the players in the market.

Regulatory instruments

Overview

A general distinction may be drawn between command-and-control regulation (which will be described as administrative regulation in this paper) and market-based regulatory instruments. Most environmental regulation has been administrative in nature, though market-based instruments have been used in the field of combustion emissions control, notably in the US 1990 Clean Air Act (CAA) Amendments.[3] There are very few examples of any individual instrument being used in isolation and real-life regulation tends to employ pragmatic combinations of the various types of control. The interacting influences of different types of instrument are often of crucial importance.

Environmental taxes

Environmental taxes are the purest form of market-based instrument, sending the 'correct' economic signals at each stage of the production process. Taxes can, in principle, influence technology choice as well as pushing up the price of final goods (e.g. electricity) and creating incentives to reduce demand. As well as economic efficiency considerations, the other potential advantage of taxes is that they provide continuing incentives for technological improvements in process plant or clean-up equipment.

Ideally, environmental taxes would be set with reference to the social and environmental costs not internalised within existing markets. However, environmental taxes which have already been introduced or are proposed tend either to be 'regulatory' (designed to achieve a given emissions objective) or fund-raising (to raise money for specific programmes).

Taxes have been little used in practice. It has proved enormously difficult to establish a sufficient degree of consensus for their introduction. From the point of view of those operating polluting processes, the most unsatisfactory aspect of environmental taxes is that residual emissions must be paid for even after the application of control technology. With all of the other regulatory instruments described, companies pay only for emissions abatement. The political difficulties being encountered by the European Community (EC) Commission's proposal for a carbon/energy tax[4] and the failure of the Clinton BTU tax in the US[5] illustrate the difficulties of implementing taxes in the face of strong industrial (and consumer) opposition. The fate of the sulphur tax concept floated by President Nixon in his 1971 State of the Union address is a further illustration of the practical problems.[6]

Where there is a strong correlation between emissions and the physical composition of fuels, a useful alternative is to tax the fuels themselves. This is the logic behind differential taxation of leaded/unleaded petrol and the EC's proposed carbon/energy tax. It could in principle be used for SO_2, but would not be appropriate for NOx emissions which depend on combustion conditions as well as fuel quality.

Externality adders

The 'externality adder' is a semi market-based instrument used by some US public utility commissions when assessing proposals for new investment in electricity supply. The adders are notionally related to the external and social and environmental costs of alternative power generation systems. The adders are used to influence the selection of new plant, or the use of demand side management (DSM), but the imputed external costs are not passed on to consumers. The main effect of this device is to promote the use of renewable energy and DSM.

Technology standards

Most pollution control regimes incorporate legally specified, qualitative standards for the types of technology which should be employed. Examples include 'Stand der Technik' (Germany), 'best available technology/technique not entailing excessive cost' (EC/UK)[7] and 'best available control technology' (US). In practice, most regulatory authorities make quantitative interpretations of these concepts, usually via emission limits in order to avoid making arbitrary decisions. Regulatory authorities which develop too specific a definition of 'best available technology' can act as a very strong brake on innovation. Where this has been the practice, companies developing innovative technologies with a high standard of performance have often found it difficult to persuade authorities to accept their solutions.[8]

Emission limits

Emission limits or standards have been the most common form of instrument used to control combustion emissions. They can take several different forms, often dictated by what is readily measurable by sampling either energy inputs or flue gases. In Europe, emission limits are related to the weight of pollutant in a given volume of flue gas (milligrammes/normal cubic metre - mg/Nm^3). The 19th Century practice of injecting cold air into the bottom of chimneys to dilute pollutants quickly led to the 'normalisation' of flue gas conditions with reference to temperature, pressure and reference gas (oxygen or carbon dioxide) concentrations. In the US, limits are set in terms of the quantity of pollutant in relation to the heat content of fuels used (pounds per million British Thermal Units (BTUs) - lbs/ Million British Thermal Units (MMBTU)). Interestingly, these formulations provide minimal incentives to increase the efficiency with which fuels are used. Relating emission limits to plant output (e.g. the kWh produced by a power station) rather than plant input would provide these incentives.

There are differences in practice in the degree to which emission limits are 'standardised'. In the US, the new source performance standard (NSPS) in lbs SO_2/MMBTU is applied across a wide range of plant types, plant sizes and fuel types. In Europe, emission limits usually depend on the type of fuel burned and the nature of the plant. Thus, larger plants facing lower unit abatement costs because of economies of scale in clean-up equipment face tighter limits than do smaller plants. Cleaner fuels, such as natural gas, face tighter standards to reflect the lower level of uncontrolled emissions. The effect of the European approach is that marginal costs of abatement across different types of plant will tend to be more equal. While these non-standardised limits were set more with reference to available technology than economic principles, they do have an economic logic.

Emission limits are most commonly applied to *new plants*. In this way, all costs can be appraised in advance by investors. In general, regulators tend also to treat significantly modified plant as if it were new for regulatory purposes. Emission limits for existing plants can be resented because they may effectively be introduced retrospectively.

The stringency of emission limits can have an important effect on markets for fuels. Modest emission limits, such as the original US NSPS of 1.2 lbs SO_2/MMBTU can be met by *some* fuels, e.g. Western strip-mined coal, without the addition of clean-up equipment to a power station. On the other hand, limits in Japan and the European Community have been set at levels which require clean-up equipment to be installed regardless of the quality of the feedstock coal.

The specification of the *averaging time* for pollution measurements is an important part of a regulation. For example, the sulphur content of coal and hence unconstrained emission levels vary over time. The shorter the averaging time, the more stringent is the control for a given emission limit. The EC's Large Combustion Plant (LCP) Directive uses a monthly averaging time but requires, in addition, that 97% of 48 hourly mean measurements are below 110% of the monthly limit.

Emission limits, notably the original US NSPS for SO_2, have been used for the purpose of 'technology forcing'. In 1971, the NSPS was set at a level which could not be achieved by any commercially available clean-up equipment.[9] The standard was intended to (and did) stimulate the commercialisation of such technologies. However, a drawback of emission limits which remain set for some time is that they provide minimal incentives for technological change. Once the standard has been achieved there is no incentive to develop technologies which improve on performance. Regulatory stability and technological considerations are in conflict in this situation. Technology may be improved if standards are tightened in line with, or in anticipation of, improvements in technological performance. However, the original US NSPS has remained essentially unchanged since it was first promulgated. The 1.2 lbs SO_2/MMBTU figure indeed re-appears in the 1990 CAA Amendments. This illustrates how difficult it can be to move standards forward.

Emission limits for existing plant

In most regimes, allowance has been made for the application of emission limits to existing as well as new plants. This has generally been the case because it has been determined at the political level that pollution loads should be reduced more quickly than the 'natural' rate of replacement of plant would allow. Several different means have been used for implementing emission limits for existing plant. In Germany, the 1983

Grossfeuerungsanlagenverordnung allowed existing power plants a specified number of operating hours before they were required to meet emission limits set for new plants.[10] The 30,000 hour residual operating time limit set was equivalent to about five years operation. In the UK, HM Inspectorate of Pollution has indicated that existing plants should meet new plant standards by a specific date, 1 April 2001, ten years after the promulgation of the regulation.[11] The EC's draft Directive on Integrated Pollution Prevention and Control also adopts the idea of a ten-year rolling programme for bringing existing plant up to new plant standards.[12] In the US, emission limits for individual existing plants have been set according to ambient air quality under state implementation plans (SIPs). In the past, these limits have often been very modest and have required, with some exceptions, adjustments to fuel supplies.

Removal requirements

The requirement to remove a specified proportion of the pollutant present in raw materials or fuels has been used in both the US and the European Community. The 1977 CAA Amendments in the US required that 90% of the sulphur be removed from coal (by pre-combustion cleaning or post-combustion clean-up) unless it could meet a relatively low standard of 0.6 lbs SO_2/MMBTU. If this standard could be met, the removal requirement was reduced to 70%. The EC's LCP Directive allows desulphurisation rates to be substituted for emission limits where indigenous solid fuel is used and 'emission limit values ... cannot be met, owing to the particular nature of the fuel, without using excessively expensive technology'. In the US, removal requirements were used in addition to a modest emission limit (equivalent to 0.7% sulphur coal) in order to ensure that low-sulphur Western coal would not gain competitive advantage. In the EC, the removal requirement is used as an alternative to a strict emission limit (equivalent to 0.2% sulphur coal) in order to permit the marketing of lower quality fuels which could not comply with the basic emission limit even after the application of best available clean-up technology.

In general, removal requirements have been used to mitigate the fuel market impacts of raw emission limits. The results have some economic logic in that they may result in a less uneven distribution of marginal abatement costs across different emission sources and fuels - i.e. the problem of zero marginal abatement costs for relatively clean fuels and very high marginal abatement costs for dirty fuels.

Fuel quality standards

Considerable use has been made of regulations governing fuel quality, e.g. lead and sulphur content, particularly in relation to petroleum products. This type of regulation falls directly on fuel suppliers rather than fuel users and is particularly appropriate for dealing with pollution arising from large numbers of smaller sources. Fiscal measures, e.g. those applied with respect to lead in petrol, are also an appropriate way of dealing with small sources.

Quotas

An alternative to the specification of emission limits is to allocate annual pollutant quotas to individual sources, or to companies operating a group of plants. An emission quota is much more flexible than an emission limit which requires continuous compliance (over the averaging time) regardless of plant output. Compliance with an emissions quota can be achieved by reducing plant output as well as by fuel selection or the installation of clean-up equipment. Where emission quotas are set on a company wide basis (company 'bubbles'), there will be even greater flexibility about compliance options. By enhancing discretion on the part of the regulated company, quotas may improve the prospect of more cost-effective compliance.

Plant specific emission quotas are used in Japan, where they are related to chimney height and ambient air quality (the k-factor system).[13] The UK national plan for SO_2 and NOx emissions establishes company-wide emission quotas for large generating companies.[14] In addition to these, HM Inspectorate of Pollution (HMIP) has also set annual plant specific emissions quotas. The plant specific quotas provide a 'safety net' in case cost-effective compliance with the company bubble would lead to excessive emissions at any individual plant. The quota system in the UK applies only to existing plant, while new plant is subject to more conventional emissions limits and removal requirements. The UK quotas are supplementary to, rather than instead of, emission limits.

Quotas by themselves make sense only when it does not matter when, within a year, emissions arise. This is unlikely to be the case if the environmental target relates to ambient air quality which responds very quickly to emission levels. However, with problems such as acid rain, where the cumulative impact of emission loads is the most important factor, quotas may be more acceptable. Company wide quotas make sense only where the specific location of emissions is not critical, i.e. in dealing with acid rain. However, even here, it is becoming possible to establish links between particular power stations and specific sensitive eco-systems.

Like emission limits, quotas do not provide specific incentives for technological change.

Tradeable permits

Tradeable emission permits are a further refinement of the concept of emission quotas, with tradeability added. Tradeable permits for air pollution are most developed in the US.[15] Although ostensibly a 'market-based instrument', the US system developed in a piecemeal fashion during the late 1970s and early 1980s in order to compensate for inflexibilities in the complex system of administrative control which had been built up through the 1970 and 1977 CAA Amendments and various court cases.[16] The original trading system allows firms to bank emission reduction credits gained through 'over-compliance' with a standard. The credits can then be used by the company itself, or sold to another company, for use in *offsetting* (allowing new plants to operate in areas where ambient air quality standards are being exceeded), *netting* (allowing significantly modified plant to escape definition as a new plant) or *bubbles* (switching emissions among a group of existing plants).

The 1990 US CAA Amendments have instituted a much more comprehensive system of emissions trading.[17] Existing power stations have been allocated emission permits based on their fuel consumption in the period 1985-87. In Phase I, by 1995, the 110 largest and dirtiest power plants are allocated permits equivalent to 2.5 lbs SO_2/MMBTU. By the year 2000, 2000 smaller power plants are allocated permits equivalent to 1.2 lbs SO_2/MMBTU (the new source performance standard). In principle, electric utilities are free to trade these emission permits in order to achieve compliance. In practice, trading between electric utilities has played little part in Phase I compliance plans although utilities have swapped permits between their own plants.[18]

In part this stems from the homogeneity of sources affected in Phase I, but is also attributable to state-level actions which have reduced incentives to trade. Unlike the emission quotas in the UK, the US tradeable permits apply to new as well as existing plants. This can result in utilities hoarding permits, with the encouragement of state public utility commissions (PUCs), in order to accommodate future growth. Some PUCs have encouraged utilities to install clean-up facilities (flue gas desulphurisation - FGD) rather than switching to lower sulphur, out-of-state coals in order to protect local jobs in the coal mining industry. More comprehensive trading may emerge in Phase II of the programme.

In general, a tradeable emission permit system will not provide incentives for improved technological performance. However, the question of technological change has been addressed within the 1990 US CAA Amendments. 'Advanced coal combustion technologies' are encouraged by allowing delayed compliance. An advanced coal combustion technology is defined as one achieving a high level of sulphur removal. The effect is to encourage technologies which will permit lower quality fuels to be used more cheaply.

Regulatory styles and market structure

Up to this point, the discussion has focused on the selection of regulatory instruments set against a background of competition between technologies and plants. However, regulation in the real world is more complex than this simple model would suggest. In particular, the *style* of regulation and the institutional framework within which it takes place may be equally (or more) important than the regulatory instruments chosen. Equally, competition, if it takes place at all, does not take place simply between technologies and plants but between companies with different plant portfolios and different technological capabilities. This other dimension also deserves attention.

Regulatory style

The regulatory process involves more than a public agency choosing a regulatory instrument and applying it to constrain the operations of polluting firms. In practice, regulations are established through negotiations in which companies themselves play a major part. In the US, the Environmental Protection Agency (EPA) is empowered by an Act of Congress to promulgate draft regulations which are then subject to wide public comment by affected companies, trade associations and environmental pressure groups before any final decision is made. The passage of the original Act will be the target of considerable lobbying effort both in Committees and on the floor of the House itself. Even after the regulations are agreed, they may be challenged in the courts. The key point is that potentially affected companies (and others) may have a major influence on the content of regulations and the way in which they are applied. They therefore help to shape responses in terms of investment and technological choice.

The three elements of regulatory style which may have implications for plant investment and technological choice are: a) the degree to which the system is legalistic or relies on administrative discretion; b) the degree of flexibility in the system; and c) the nature of the relationship between the regulatory agency and industry. If regulatory systems are legalistic in nature, then there is little room for negotiation between regulators and industry at the implementation stage. If, in addition, the process of developing new laws is slow and cumbersome, as in the very pluralistic political environment of the US, then standards may become embedded for a very long time, diminishing incentives for technological change. The development of emissions trading in the US can be viewed as a response to an initially inflexible regulatory/legal system.

When faced with lengthy rule-making processes, there are incentives for regulators to take account of the views of industry, formally or semi-formally, as early in the process as possible. This approach is explicitly endorsed by the European Community in its new Action Programme for the Environment.[19] The close relationships which may develop between the regulator and the regulated can result in a particularly strong industrial influence over regulations. The degree to which this promotes or inhibits investment and technological change must depend on individual circumstances and the interests of the key players.

In practice then, the selection of regulatory instruments is not simply the pursuit of the public interest, but may be heavily constrained by the nature of the regulatory system and the interests of regulated companies.

The nature of the regulated market

The 'model' of investment described earlier said nothing about the type of market within which different plants and technologies might be employed. It could be applied to a single body, such as a monopoly electricity supplier, selflessly pursuing a cost minimisation objective. Equally ideally, it could be applied to a perfect market in which cost minimisation was achieved through competition. However, in the real world, markets and firms may operate as imperfectly as regulatory agencies. The structure of markets and the parallel regulation of prices and profits matter and will have important effects on investment and technology choices. For example, in most parts of the world, electric utilities operating as monopolies are subject to rate-of-return regulation which may blunt incentives to select the lowest cost methods of compliance with regulations. This raises the important question of multi-regulation and the relationship between regulatory bodies operating in the environmental and economic domains.

In some markets, there may be significant imbalances between the plant/technology portfolios of different players in the market. Under these circumstances (e.g. in the UK electricity market), environmental regulations may have important impacts on market entry (and exit) as well as on the relative profitability of the different players. As with different regulatory styles, the impacts on investment and technological change are dependent on the particular circumstances in particular markets.

Regulatory issues and new investment

Plant replacement rates

It has been noted that the 'natural' replacement of plant provides an appropriate opportunity to embody higher standards of environmental protection. An environmental tax could be set at a level which would induce the application of cleaner technology in new plants while maintaining approximately the same rate of plant replacement. However, while this may be a desirable *outcome* (if the implied rate of emissions reductions is politically acceptable), an administrative instrument (e.g. emission limit) which simply requires that new investment must operate at a higher environmental standard is unlikely to have the desired effect.

A hypothetical environmental tax can be used as a benchmark against which to judge the appropriateness and efficiency of other instruments. An environmental tax could be set at a level such that it would push up the cost of both new and existing plants, leaving the basic decision about whether to replace a plant or not little affected. However, administrative instruments applied only to new plants serve to make new plants more expensive to build and operate, while leaving the cost of operating existing plants unaffected. Plant operators will have incentives to adjust maintenance regimes so that existing plants are retained in service for longer periods. In the short term, the application of administrative measures to new plants only, as was the case during the 1970s in countries such as Germany and the US, may have perverse effects in that investment in cleaner, more efficient plant is delayed.

In practice, this problem has probably diminished because emission controls have now been applied to existing plants in order to generate a rapid rate of emissions reduction. However, the application of emission controls to existing plants appears desirable on investment as well as environmental grounds.

Existing plant controls and lock-in to old technologies

An environmental tax set at a level sufficiently high could be used to cause emissions to fall at a rate more rapid than that resulting from natural plant replacement. It would have the following effects:

- it would cause some older, less efficient and more polluting plant to be retired early because its cost would be pushed up relative to new plant and the costs of retrofits would need to be written off over an unacceptably short period;
- it would stimulate the modification of some newer and more efficient existing plant through the retrofitting of clean-up equipment where the residual operating life was sufficiently long; and
- it could stimulate a switch to less polluting but more expensive fuels at intermediate plants where neither closure not expensive retrofitting was justified.

Establishing a deadline for existing plants to comply with emission limits would achieve approximately this result. The two key questions are: a) the stringency of emission limits with respect to those for similar new plant; and b) the length of the compliance schedule. Where retrofitting/closure are likely to be the dominant options, then setting existing plant standards at the new plant level is reasonable. Switching to premium fuels where this results in emission levels higher than those from new plants suggests, however, a less stringent emission standard for some categories of plant.

Ideally, the compliance deadline would be no longer than the period required to plan and execute plant modifications and closures. This is the response which the introduction of an environmental tax would induce. In practice, a longer period might be preferable depending on the capacity of the industry to phase and fund investment. Although power plants for example may have a 30-40 year lifetime, there is generally a much shorter cycle of planned maintenance, of the order of two-three years, with which a retrofitting programme might be phased. The length of the compliance schedule might also be influenced by the maturity of the retrofit technology. If a technology is relatively new, then there may be a rationale for extending compliance deadlines so that experience can be built up gradually and knowledge gained from early plants can be used in later projects.

The German programme of retrofitting FGD and selective catalytic reduction (SCR) technology to existing power stations in the 1980s and early 1990s reflects this pattern.[20] FGD is a mature technology for abating SO_2 emissions. There was a five year deadline for SO_2 compliance which was as fast as technically possible. SCR for abating NOx was much less developed and Japanese technology had to be adapted to the German situation. Here a longer compliance schedule of approximately ten years was adopted.

The problem with the German solution lies in its robustness with respect to technical change in power generation. Since the German FGD programme was completed, the prospect of substantially cleaner and cheaper power generation in combined cycle gas turbine (CCGT) stations has emerged through a combination of technological change, resource availability and the lifting of legal restrictions on gas burning. With hindsight, it is not clear whether the actual investment programme adopted was the most cost-effective. Had a longer compliance schedule been adopted, of the order of 10-15 years, then it would have been possible for new technology to have been introduced more rapidly at a later date. This problem of 'lock-in' to existing technology is not necessarily resolvable through the choice of regulatory instrument. The problem arises because no instrument, whether market-based or administrative in nature, can be completely robust against rapid or substantial changes in market conditions or the availability of technology. However, more gradual compliance schedules will allow industry to hedge its bets against less predictable changes in markets and/or technology.

Migration of polluting activity

In general, environmental controls tend to be focused, at least initially, on the largest individual sources of pollution, such as power stations. This reflects an understandable desire to obtain higher environmental improvements with the lowest direct and administrative costs. However, this can lead to perverse effects if the problem of smaller sources is entirely ignored. For example, the EC's LCP Directive, which has a lower size threshold for plant, specifies emission limits for medium-sized plant which can be met through the use of premium low sulphur fuels. By itself, this will eventually result in the use of higher sulphur fuels migrating to plant which falls below the LCP threshold. Not only will this diminish the effectiveness of the Directive in global terms, it will also result in higher sulphur fuels being burned in less appropriate situations, i.e. at plant with lower chimneys located in urban areas.

There are clear administrative difficulties associated with applying a system of emission limits to very small plant because of inspection and enforcement difficulties. More pervasive instruments, such as fuel quality standards or even environmental taxes, would be suitable for addressing this problem.

Robustness to market/technological change

In the case of the German FGD problem, the problem of 'locking-in' to obsolete technologies through short compliance schedules was raised. In part, the British method of allocating declining emission quotas to either large companies, or sectors comprising smaller companies, can address this problem. If the quotas decline over a sufficiently long period, there is flexibility to modify the abatement strategy to take account of technical change and altering market conditions. The flexibility accorded to individual companies allows non-uniform and more cost-effective responses at individual plants.

However, a perversity again arises if the level of emission quotas established is overtaken by changing market circumstances. The UK emission quotas were negotiated before there was an expectation (at least on the part of Government) that CCGT stations would displace large quantities of coal from the British electricity supply system. As a result, unconstrained emissions are now substantially below the quota levels and incentives even to operate stations fitted with FGD have been blunted. Abatement opportunities with a very low marginal cost (running a station fitted with FGD as opposed to running an uncontrolled one) may be passed up. If the emissions quota reflects a threshold below which environmental damage does not occur (the critical loads approach), then the low marginal cost of abatement may be acceptable. However, if emissions are well above such thresholds, as is the case for the British electricity supply industry, then the outcome is clearly perverse. This is another example of how a regulatory instrument may not be robust to changing circumstances. This conclusion would also apply if the emission quotas were tradeable. Whatever instrument is chosen, flexibility is required to deal with substantial changes in the business environment.

Asymmetries of information

The problem of establishing an appropriate level of control where the industry is better informed than the regulatory agency is a familiar problem.[21] The informational asymmetry can occur with respect to both knowledge of available technologies and knowledge of broader market trends. Although the former asymmetry has been most widely discussed, it is likely that the degree of asymmetry is greatest with respect to knowledge of wider market issues. A reasonably accurate assessment of the cost of a given abatement technology (e.g. FGD or SCR) can probably be gained by the regulator at a fairly low cost. However, market or company circumstances may be more difficult to appreciate, particularly if the culture of the regulatory agency is engineering-based.

Again, the British emission quotas provide a useful example. At the time they were negotiated between Government and the electricity supply industry, substantial changes in both the structure of the market and the choice of technology for new generation was in the offing. The Government and HMIP almost certainly did not appreciate the extent of the changes which were imminent. Had the electricity supply industry been able to foretell these changes, even partly, it would have been in a position to negotiate emission quotas which were higher, and required less active abatement measures at a lower marginal cost, then would otherwise have been the case.

Market structure and regulation

Environmental regulation can have significant differential effects on participants in a market. In principle, environmental taxes are most equitable in that they would allow different participants to compete on a level playing field. However, that is not to say that individual companies operating older, more polluting plant would not be disadvantaged.

The problem of market structure and environmental regulation is brought into sharpest perspective where permit/quota based systems, whether embodying tradability or not, have been used. The problem is that the process of 'grandfathering' out permits or quotas is equivalent to allocating a property right. These can be a powerful instrument of competitive advantage. In the US, the tradeable permit system covers both new and existing plant. By hoarding permits, strong existing players in the market can inhibit entry and retain a dominant market position, probably inhibiting new investment and technical change in the process. This problem has been acknowledged and at least partially addressed through the auction of permits earmarked for new entrants.

In the UK electricity market, non-tradable emission quotas have been allocated only to existing plant, with new plant subject to emission limits. Following electricity privatisation, existing plant in England and Wales is owned exclusively by two large generators. The current balance of plant economics suggests that the appropriate response to any tightening of the existing plant quotas in the medium-long term will be plant closure rather than retrofitting. Hence, the question of environmental controls is intimately linked to the rate of technical change, the type and level of investment and competitive positioning in the market. This situation calls for a link between environmental regulation and economic oversight of the market.

Clean-up, cleaner technologies and long-term technological change

A great deal of recent discussion has focused on the desirability of promoting 'cleaner technology' (intrinsically cleaner processes which may also be cheaper) as opposed to 'clean-up technology' (end-of-pipe solutions which inevitably increase costs and may simply transfer pollution to different media).[22] CCGTs and advanced coal combustion technologies might be thought of as 'cleaner', while FGD is certainly a 'clean-up' technology.

More than twenty years after the introduction of SO_2 controls on power stations and other combustion installations, FGD remains the dominant system for reducing emissions.[23] In spite of RD&D which has taken place since the 1970s, advanced coal combustion technologies such as pressurised fluidised bed combustion (PFBC) and integrated gasification combined cycle (IGCC) have barely been commercialised.[24] What lessons can be drawn about environmental regulation and longer-term technological change?

First, it may be difficult to develop radical cleaner technologies on a sufficiently short timescale to meet political demands for reduced emission levels. This means that simpler clean-up technologies such as FGD, based on well understood physical and chemical principles, may be required to meet actual demands for emission reductions. Second, however, emission regulations and the *expectation* of tightening standards in the longer term have pushed up the cost of conventional power generation systems and have created incentives to develop advanced technologies which are potentially cheaper as well as cleaner. Thus, clean-up technologies may, ultimately, act as a bridge between the uncontrolled operation of traditional technology and the ultimate adoption of cleaner techniques. Third, is the role of RD&D activity in developing cleaner techniques. Private companies such as ABB, Shell and Texaco have made substantial investments in advanced coal technology. However, particularly in the US there has been a substantial level of public RD&D support which has served to promote advanced coal technologies. To some extent, the RD&D activity has been supported as part of a *quid pro quo* for the acceptance of tighter environmental standards by powerful coal industry lobbies. As well as emission standards providing a push for private sector involvement in RD&D, they have also served to stimulate activity in the public sector.

Finally, it is worthwhile observing the very rapid adoption of clean CCGT technology in the UK where the availability of technology was matched by a market need. The availability of this technology is largely due to techniques and new materials (more efficient and durable gas turbines) being 'parachuted' into the electricity sector from other areas of technology development (e.g. aviation). It remains to be seen whether the adoption of the generic techniques and materials embodied in CCGTs will have a more important longer term impact than 'forced' technology such as advanced coal. Perhaps IGCC might represent some marriage of the two separate sets of developments.

Conclusions

The diversity of circumstances, regulatory styles and choice of instruments in various countries makes it very difficult to draw any generalised conclusions about the linkages between environmental regulation, investment and technological change. However, the following preliminary conclusions are advanced:

- the choice of regulatory instruments to control emissions can have profound effects on patterns of plant investment (level and composition) and on incentives to adopt more novel technological solutions;
- the choice of regulatory instruments is conditioned by the style of regulation used in any particular country;
- the energy industry itself has an important role to play in shaping regulatory solutions. Environmental controls which are adopted often reflect the needs and technological capabilities of regulated firms;

- there are many examples of environmental controls leading to perverse results - inhibiting the adoption of cleaner techniques, generating weak incentives even to use installed clean-up technology. Sometimes this is due to poor regulatory design (e.g. inflexible interpretations of principles such as 'best available technology'). On other occasions it is simply because no system of regulation can keep pace with substantial changes in market conditions or technology availability;
- it is important that industry is involved in the development of regulations so that realistic controls which reflect technological capabilities are developed. On the other hand, there needs to be sufficient flexibility in the regulatory process to refine controls so that changes in markets and technology availability can be reflected and that industry does not benefit unduly from informational asymmetries; and
- RD&D activity (the technology push) has a role to play as well as regulation (the demand pull) in developing radical and cleaner solution to environmental problems. In some cases, the higher cost of conventional technology coupled with expectations of tighter controls in the long term and the desire of innovative companies to gain competitive advantage may be sufficient to move technology forwards. In other cases, there may be a need for publicly supported RD&D activity.

Acknowledgements

The author gratefully acknowledges ever-helpful comments on an earlier draft of this paper from Professor John Surrey of the Science Policy Research Unit at the University of Sussex. The paper is based on research supported by: British Gas and the Economic and Social Research Council through a Professorial Fellowship on the Adoption and Development of Clean Technology in Industry; and by the Joint Committee of the Economic and Social Research Council and the Science and Engineering Research Council for work on *Environmental Regulation, Technical Change and the Energy Sector* (Grant GR/E /70405). However, the views expressed in the paper are those of the author alone.

Notes

1 WEG Salter, *Productivity and Technical Change*, Department of Applied Economics Monograph, University of Cambridge, 1969

2 see for example, Organisation for Economic Cooperation and Development, *Economic Instruments for Environmental Protection*, OECD, Paris, 1989

3 US Environmental Protection Agency, *The Clean Air Act Amendments of 1990: Summary Materials*, Washington DC, November 1990

4 Commission of the European Communities, *Proposal for a Council directive introducing a tax on carbon dioxide emissions and energy*, COM(92) 226 final, Brussels, 30 June 1992

5 JE Treat and AE Sieminski, *US Energy Tax: Winners and Losers*, Washington Analysis, Washington DC, March 1993

6 A Myrick Freeman, RH Haveman and AV Kneese, *The Economics of Environmental Policy*, Wiley, New York, 1973, p5

7 The European Community is proposing to move over to an integrated pollution control system based simply on the *best available technology* principle. See the explanatory memorandum in: Commission of the European Communities, *Proposal for a Council Directive on Integrated Pollution Prevention and Control*, COM(93) 423 final, Brussels, 14 September 1993

8 Organisation for Economic Cooperation and Development, *Environmental Policy and Technical Change*, OECD, Paris, 1985

9 BA Ackerman and WT Hassler, *Clean Coal/Dirty Air*, Yale University Press, 1981

10 July 1983, 'Grossfeuerungsanlagenverordnung' (GFAVo), *13th Bundesimmissionschutzverordnung*, Bonn

11 HM Inspectorate of Pollution, *Combustion Processes: Large Boilers and Furnaces 50MW(th) and Over*, Process Guidance Note IPR 1/1, HMSO, London, 1991

12 Commission of the European Communities, *Proposal for a Council Directive on Integrated Pollution Prevention and Control*, COM(93) 423 final, Brussels, 14 September 1993

13 H Nishimura, *How to Conquer Air Pollution: A Japanese Experience*, Studies in Environmental Science 38, Elsevier, Amsterdam, 1989

14 Department of the Environment, *The UK's Programme and National Plan for Reducing Emissions of SO2 and NOX from Existing Large Combustion Plants*, London, 20 December 1990

15 *THE Tietenberg, Emissions Trading: An Exercise in Reforming Pollution Policy, Resources for the Future*, Washington DC, 1985

16 *op cit*

17 US Environmental Protection Agency, *The Clean Air Act Amendments of 1990: Summary Materials*, Washington DC, November 1990

18 'EPA Holds Allowance Auction, Board of Trade Plans Private Sale', *Public Utilities Fortnightly*, Vol 131, No 9, 1 May 1993, pp38-39

19 Commission of the European Communities, *Towards Sustainability: A European Community Programme of Policy and Action in relation to the Environment and Sustainable Development,* COM(92) 23 final, Brussels, 27 March 1992

20 SA Boehmer-Christiansen and J Skea, *Acid Politics: Environmental and Energy Policies in Britain and Germany*, Belhaven, London, 1991

21 E Romstad and O Bergland, 'Inducing Individual Firm Compliance to Emission Quotas when Abatement Costs are Private Knowledge', Paper presented at the *Fourth Annual Conference of the European Association of Environmental and Resource Economists*, INSEAD, Fontainebleau, 30 June - 3 July 1993

22 T Jackson (edited), *Clean Production Strategies*, Stockholm Environment Institute/Lewis, Boca Raton, 1993

23 JL Vernon and HN Soud, *FGD Installations on Coal-Fired Plants*, IEACR/22, IEA Coal Research, London, April 1990

24 WC Patterson, *Coal-Use Technology in a Changing Environment: The Advance Continues*, FTBI, London, 1990

CHAPTER 11

Environmental protection through energy conservation: A 'free lunch' at last ?

Larry E Ruff

Introduction

Energy conservation will surely play a large role in the costly effort necessary to reduce environmental problems. Furthermore, utilities (among others) can facilitate conservation by 'reducing market imperfections,' i.e., by finding innovative ways to offer conservation services in the market in competition with energy itself. But the current enthusiasm for utility demand-side programmes goes beyond the common-sense observation that conservation will thrive if given the opportunity to compete with supply as a way to meet consumer needs at minimum economic and environmental cost. Instead, it is commonly argued that, if only normal market concepts and processes are set aside where conservation is concerned, both energy use and energy costs can be reduced simultaneously, yielding environmental benefits as virtually a 'free lunch.'

This paper examines the view that utility demand-side programmes offer an environmental free lunch on a large scale. It is shown that energy conservation neither requires nor justifies setting aside standard economic and business principles or processes. It is also shown that the more enthusiastic projections of the 'cost-effective' conservation potential are based on advocates' hypothetical claims about costs and benefits that are not supported by the available market evidence. This helps explain why calls for aggressive energy conservation efforts are directed almost exclusively at electric utilities: only administrative regulatory processes allow advocates' claims to replace valid market tests; and only a protected monopoly has the power to 'tax' its captive customers to cross-subsidise demand-side options that cannot compete in a competitive market because they are not truly cost-effective.

Of course, if energy is under priced because of uninternalised environmental costs, there is a theoretical argument for also under pricing conservation services as a 'second-best' alternative to raising energy prices. However, if environmental costs are reasonably internalised on the supply-side where most of them occur, using emission charges, marketable discharge permits, or economically-based regulations, the resulting price increases will stimulate the proper pattern and amount of energy conservation. Simple numerical examples suggest that the appropriate price-correcting subsidy is likely to be small compared to the level of subsidy implicit in many demand-side programmes, and that most of the cost-effective environmental benefit is likely to come from supply-side rather than demand-side responses; this may be true even for pollutants such as carbon dioxide for which supply-side controls are most difficult.

The concept of bundled end-use services

Kilowatt-hours (kWh) are never eaten raw, but are combined with other things - insulation, light fixtures, machinery - to produce the end-use energy services consumers ultimately desire - heat, light, mechanical drive. Thus, at least conceptually, a utility should be viewed not simply as a seller of commodity kWh, but as a potential seller of 'bundles' of kWh and complementary services, designed to meet consumers' demands for energy services at the lowest possible total cost.

In this view, utility provision of demand-side services is simply a downstream value-adding strategy. Such value-adding or bundling strategies are essential to competitive survival in most markets and have been the key to the growth of utilities in the past; vertically integrated, regional utilities bundling generation, transmission and distribution into a higher-value service called 'reliable, delivered electricity' have prospered because consumers found this bundled service good value relative to non-electric energy or self-generation.

A utility should constantly re-evaluate the service bundles it offers to determine whether different combinations might better serve its customers' end needs. It should be prepared to unbundle its services and even to reduce the scope of its internal activities if this appears competitively advantageous. For example, a utility with a particular advantage in transmission because of experience or location should consider offering transmission as an unbundled service; and a utility that can buy electricity from an independent source more cheaply than it can produce it should do so. Eventually, utilities and even consumers may have the option of buying generation, transmission and distribution from separate vendors and assembling their own bundles of reliable, delivered electricity.

By the same reasoning, a utility should expand its activities to offer bundles including additional services if doing so allows it to offer customers better value for their money. A utility with a cost advantage in mining coal should integrate upstream into this business, providing its customers with a bundle that includes utility-mined coal. A utility with an advantage in bundling its commodity electricity with other services to provide end-use energy services should integrate further downstream into consumer markets. There is nothing sacred about the traditional utility service bundles.

The ultimate level playing field

As a utility seeks higher-value services to offer its customers, it should not discriminate against service bundles that meet customers' needs with less electricity and more conservation. However, leaving environmental issues aside for the moment, neither should it discriminate **in favour of** conservation-intensive bundles, **just because** such bundles contain less commodity electricity. Any utility activity, whether downstream integration into high-efficiency lighting, upstream integration into coal mining, horizontal integration through merger with a neighbouring utility, or even diversification into an unrelated activity such as baking pizzas, might lower customer costs and enhance utility competitiveness. If it does, there is no reason in principle why a utility should not offer this service, whether it results in more or less commodity electricity being included in the utility's bundle.

This is the ultimate 'level playing field': all activities that have economic efficiency as their principal rationale should compete on their ability to lower the costs of providing consumers with the end-use services they desire. As discussed presently, some levelling of the playing field at the point of consumption may be appropriate in principle if environmental costs are not reasonably internalised where they occur. However, trying to manage supply-side environmental problems with demand-side programmes is relatively ineffective and inefficient, compared to appropriate supply-side policies that may have surprisingly little demand-side effect. Thus, environmental issues can be set aside for the moment, to be revisited in a concluding section.

Not only do all efficiency-seeking activities have the same fundamental objective of lowering consumers' end-use costs, but all such activities pursue this objective with the same fundamental strategy: reducing information and transaction costs. It is common, at least where utility demand-side programmes are concerned, to attribute information and transaction costs to something called 'market imperfections.' But the reduction of such costs, whatever their cause, is the **raison d'être** of any organised economic activity. No activity, whether inside or outside a utility, deserves special consideration just because it seeks to reduce information and transaction costs.

Consumers in all times and places meet their end-use needs by learning about technical options, finding people who will lend them resources, obtaining physical inputs and organising co-operative efforts to assemble the various pieces - in short, by obtaining information and engaging in transactions, whether to create end-use energy services or pizza services. These universal activities always involve time and inconvenience - i.e., markets are always 'imperfect' - and hence the cost of organising delivery of a service can often be lowered by an enterprise that specialises in that job. Such an enterprise is normally called a business, and its success at reducing information and transaction costs is normally measured by its ability to sell its services to willing buyers for enough to cover its (allegedly lower) costs.

A utility conservation programme tries to lower the information and transaction costs involved in delivering end-use services to customers. But precisely the same is true of a utility generation programme, a utility coal mining programme, a utility pizza baking programme or a non-utility programme. Each such economic activity, whether inside or outside a utility, and whether on the demand side or the supply side, should be required to justify its existence, not by pointing to the universal truth that information and transactions costs are **high**, but by demonstrating its ability actually to **reduce** these costs.

Utility downstream integration as an efficiency measure ?

Viewing a utility as a provider of end-use services rather than just commodity electricity provides a logical way to justify demand-side programmes on standard economic and business grounds. However, this even-handed view of demand-side programmes raises an immediate question: what reason is there to think that utilities have the kind of competitive advantage in end-use markets that would justify the more enthusiastic claims about conservation potential?

On the face of it, it is hard to see why utilities should have much comparative advantage in providing bundled end-use services to consumers. For one thing, downstream integration of utilities into dynamic, service-oriented consumer markets would run directly counter to the unbundling and vertical disintegration movements sweeping modern economies. In telecommunications, transport, finance, computers, even on the supply side of the electricity business itself, the clear trend of modern technology and institutions is toward unbundling and entrepreneurial competition in the market. What reason is there to think that large, monopolistic, regulated utilities have significant cost advantages in providing tailored demand-side services to millions of individual consumers, when so many utilities are having trouble competing even in the large-scale, engineering-oriented generation businesses they have been in for decades, and when so many utility diversification efforts have proven disappointing or worse?

Of course, utilities can learn new businesses and can restructure themselves along more entrepreneurial lines. Some of the most successful firms in dynamic, service-oriented industries are large; AT&T, American Airlines, Citibank and IBM are surviving, albeit not without difficulty in the face of increased competition and unbundling. But these large companies do not try to do everything, and what they do is done largely through decentralised, specialised profit centres that either pay their own way as separate businesses or are closed down or spun off. Boeing is very good at making aeroplanes, but lets American Airlines bundle these with other services for millions of consumers; IBM makes good business machines, but sold off its retail stores because other firms were better at bundling this commodity with end-use services for small consumers; and AT&T's computer business may benefit from being associated with AT&T's telecommunications business, but had better pay its own way or it will be history.

A utility demand-side business that is really serving consumers should be able to buy commodity electricity on an arm's-length basis from the generation side of the business or from another supplier, buy complementary goods and services in the market, and sell bundled end-use services in competition with all the other sources of such end-use services. But evidence from energy service markets does not support the view that such activities would accomplish dramatic results. There is no shortage of firms selling high-efficiency lights, insulation and advice; these firms are constantly looking for new ways to offer bundled services that consumers value. What reason is there to think that these normal, competitive market processes that work so well in providing most other end-use services are barely scratching the surface of the potential end-use energy service market?

The basis for claims about conservation potential

If there really is a large, untapped potential for cost-effective utility demand-side programmes, it must be due to some combination of two factors: either commodity electricity is badly under priced, so that consumers have too little incentive to buy cost-effective conservation services from utilities or anybody else; or utilities have some under-appreciated cost advantages that allow them to thrive in markets where ABC Conservation Company cannot. If neither of these is true, then serious consideration must be given to the possibility that the definition of the 'cost-effective potential' is badly distorted where utility conservation programmes are concerned.

The possibility that electricity is under priced because of environmental costs and the implications for a utility demand-side business are discussed presently. Electricity under pricing for other reasons was a problem following the oil price shocks of the 1970s and persists today in special circumstances, particularly in regions with subsidised hydropower and during peak demand periods in many regions. Of course, even if prices are providing inadequate conservation incentives, the logical solution is to correct the price signals, perhaps using focused subsidies where necessary, to give consumers the right incentives to conserve in everything they do; it is quite illogical to leave electricity under priced so that only a few conservation services provided by the utility can compete with it, and then only if they, too, are under priced. But under priced electricity is not a widespread problem today and is not the most common basis for claims that there is a huge, unexploited potential for cost-effective conservation.

The more common defence of implausible claims about conservation potential is that utilities have large advantages in reducing information and transaction costs caused by 'market imperfections.' The argument is that a utility demand-side business has more credibility with customers, more ability to study the options and hire the experts, better access to capital, better service personnel, more comprehensive databases and billing systems, etc., than does ABC Company, and hence can provide conservation services with much lower transaction and information costs. Thus, a utility's demand-side business should be able to deliver conservation services at prices that ABC Company cannot be expected to match.

The only problem with this argument is that few people, and least of all demand-side advocates, really believe it. The ability of a utility to reduce market imperfections is seldom cited to explain why the utility's demand-side business is making money selling better services at lower prices, thereby beating ABC Company at its own game. Instead, market imperfections are invariably cited as excuses for exempting utility demand-side programmes from the need to compete on price and service. The typical claim is that the experts know which conservation services are cost-effective for which consumers in which applications at which times, and hence there is no need to define, measure and price utility demand-side services, or to convince individual customers that the benefits are worth the costs; instead, the utility demand-side business should do what ABC Company cannot do - give away its services where the experts claim they are cost-effective and raise the price of electricity to pay for them.

This argument is sometimes stated as a research proposition, along the following lines: hypothetical arguments will never determine how large the cost-effective conservation potential really is, so let us run a market experiment; however, because we cannot sell utility conservation services for enough to cover their costs, we should give them away, paying for them with the revenue from higher electricity prices; if we can give away enough of these costly services on an experimental basis, we will expand the give-away programme, on the grounds that we are somehow lowering customer costs and enhancing utility competitiveness - even though customer electricity prices are higher and/or utility profits are lower. Such market research designs and marketing strategies would not last long in the insurance or automobile business, or even on the supply side of the utility business; but they appear to be taken seriously where utility demand-side programmes are concerned.

Now there is no doubt that utility conservation programmes can be expanded to virtually any extent if the level playing field is interpreted in this way. The experts and advocates will never run out of services they can 'prove' to be cost-effective based on their own definitions and estimates, as long as they do not have to suffer the cost and indignity of dealing with consumers as consenting adults and defining, measuring and pricing utility services. But using the utility's monopoly power in the commodity electricity market to escape competitive disciplines in downstream markets should never be confused with reducing market imperfections; indeed, it is a sure way to create new market imperfections. As long as demand-side advocates are so unwilling to subject their claims about cost-effectiveness to any reasonable market test - and trying to give away valuable and costly services hardly qualifies as a valid market test - their claims are at best unproven.

Since the more enthusiastic estimates of the cost-effective conservation potential are not based on claims that electricity is under priced or that utility programmes can lower costs enough to compete successfully in the market, one can be excused a heretical thought: **Maybe the alleged potential isn't there.** Maybe there is a reason that special concepts, criteria and jargon have had to be invented to justify demand-side programmes. Maybe, if low energy lighting, low flow showers and water heater wrapping programmes were subjected to the same tests routinely imposed on utility generation, coal mining or pizza baking programmes, the same common-sense conclusion would be reached in each case: a utility with good management and entrepreneurial abilities should be able to compete in any of these markets - but only as one of many competent players, not with such an overwhelming advantage that it is likely to change fundamentally the competitive landscape or general consumption patterns.

The 'no-cross-subsidy' test

The utility demand-side movement has spawned a legion of special concepts and 'tests' - the 'no-losers test,' the 'all-ratepayers test,' the 'non-participants test' - all for the purpose of answering one simple question: how can it be determined whether some utility programme will benefit consumers, on balance? But there is nothing new or difficult about this question for which generations of academics, managers and regulators have developed a very logical and elegant answer: in a market economy, an efficient activity is one that results in lower average prices - not lower revenues or customer bills of a specific enterprise, but lower **prices**.

There is no more fundamental economic proposition than this - that an economic enterprise should be judged by its ability to keep its unit costs, and hence its prices, low. In keeping with demand-side tradition, this proposition can be stated as a 'test' - the **no-cross-subsidy test**. According to the no-cross-subsidy test, a utility should not engage in any activity if doing so results in average prices for other utility services being higher than they would otherwise be. Although this test has appeal on distributional or equity grounds, it is strictly an efficiency test based on very simple logic: if those who benefit from a utility programme are not able to pay its incremental costs and still be ahead of the game, then the programme must have more costs than benefits and is not cost-effective by any reasonable definition of that term.

The most natural and effective way to satisfy the no-cross-subsidy test is to require the costs of each service to be recovered through the prices charged for the service, so that alleged beneficiaries have the option of declining to pay for the service if its costs exceed its benefits. However, in principle the no-cross-subsidy test can also be met by recovering the costs of a class-wide service through a surcharge on customers in the class. This method of cost recovery might be applicable in special cases where some simple service is so obviously beneficial for all customers in a class that it can simply be delivered and its costs collected through a surcharge or tax on the lucky beneficiaries - an approach that at least gives the class of alleged beneficiaries an incentive, and presumably an opportunity through the regulatory process, to object if they do not think the benefits are worth the costs.

The no-cross-subsidy test, under different names, is routinely applied to utility supply-side programmes, despite persistent confusion on this matter. When demand is growing, a utility with an obligation to serve, has no choice but to meet that demand, even if doing so increases prices for the 'old' customers whose demand is not growing. The no-cross-subsidy test says that the best option in such cases is the one that results in the **least** increase in electricity prices. When only supply-side options are being compared, the demand to be met is the same for all options, and hence the option that minimises the present value of revenue requirements also minimises average prices - since REVENUE = PRICE x QUANTITY, and QUANTITY is the same for all supply-side options, minimising REVENUE is equivalent to minimising PRICE. In this special case, the no-cross-subsidy test is equivalent (apart from rate-design issues that are not directly relevant here) to the traditional minimise-revenue-requirements test.

For evaluating utility options more generally, the no-cross-subsidy test remains valid while the minimise-revenue-requirements test is clearly inappropriate. If a utility were to propose adding coal mining or pizza baking to its activities, there would be no fundamental reason to object, as long as engaging in these activities reduced average **prices**. If getting into the pizza business allowed the utility to lower its electricity prices, nobody would object that the total utility revenue requirement or utility bill, now covering both electricity and pizza, is higher. But if the only way the utility can pay for mining coal or baking pizza is to charge higher prices for electricity, consumers and regulators - and probably even most demand-side advocates - would protest vehemently. They would say that, while utility integration into such businesses is not objectionable in principle, if it must be cross-subsidised through higher electricity rates it must be inefficient and should not be undertaken by the utility.

Even though there is no objection in principle to a utility getting into coal mining or pizza baking as long as no cross-subsidy is involved, in practice regulators are understandably reluctant to allow utility diversification outside the core electricity business, simply because they are **afraid** of cross-subsidisation. They fear that accounting cost allocations and other difficult-to-monitor devices will be used to raise the price of the commodity electricity all customers buy, with the proceeds used to cross-subsidise utility activities that cannot otherwise compete. Nor are they much swayed by the fact that the subsidised services are on the demand side (e.g., appliance sales), are in markets that are highly imperfect, are cost-effective for many customers, or involve 'only a transfer payment' from ratepayers who choose not to use the subsidised services to those who do. They know, as surely as they know anything, that even running the risk of such cross-subsidies is bad economics, bad business and bad regulatory policy.

Despite this deep and proper aversion to cross-subsidising utility programmes in general, for some reason different standards have become accepted where demand-side programmes are concerned. Of course, nobody advocating a demand-side programme is foolish enough to say that it is cost-effective even though it must be **cross-subsidised**; instead, they say that it is cost-effective even though it cannot pass a 'no-losers test' or a 'rate-impact test.' But, under this rhetorical smoke screen, they are quite explicitly arguing that utilities should be allowed, even forced, to raise the price of commodity electricity to pay for utility entry into markets where the utility cannot otherwise compete. If this is not a cross-subsidy, what is? And if this is the basis for the claims about the massive potential for 'cost-effective' utility demand-side programmes, why should these claims be taken seriously?

Why utility conservation and not utility pizzas ?

The widespread view that demand-side programmes are somehow different, and hence should be exempt from standard economic and business criteria such as the no-cross-subsidy principle, has absolutely no basis in economic or administrative theory. Indeed, the entire intellectual foundation for this view comes from simple examples purporting to prove that applying a no-cross-subsidy/no-losers/no-rate-increase test discriminates against demand-side programmes. Yet, hidden in these examples is a critical, implicit assumption that would be immediately spotted and rejected if applied to, say, utility pizza programmes.

A typical example allegedly 'proving' the inapplicability of a no-cross-subsidy test for conservation programmes goes as follows: a utility is selling electricity for its marginal supply cost of 8 cents/kWh; demand-side advocates say that a high-efficiency light can save a kWh for only 5 cents at no cost to customers in terms of quality or convenience of service - **if** the utility 'reduces market imperfections' by installing the light itself; however, if the utility provides the light rather than the generated kWh, it will have higher total costs with no increase in kWh sales and will have to increase average electricity prices; thus, applying a no-cross-subsidy test would prevent the utility from implementing an obviously cost-effective demand-side programme.

WHAT IS WRONG WITH THIS PICTURE ?? Most people tend to nod in agreement at each step in this argument; if the conclusion appears unusual, they apparently assume that it has something to do with the special energy-saving or market-improving character of demand-side programmes. After all, if a utility pays for a conservation measure that only it can implement, even a very cost-effective one, of course it will have higher costs with smaller electricity sales and will have to raise average electricity prices to pay for it. But, in fact, the conclusion in this example has nothing to do with any special characteristics of utility conservation programmes and everything to do with an implicit assumption about utility pricing. What is it?

Give up? As a hint, try applying precisely the same logic to a utility pizza programme: pizzas cost $8 in the market; utility pizza advocates say that pizzas just as tasty and nutritious can be made for only $5 - **if** the utility 'reduces market imperfections' by making the pizzas itself; however, if the utility produces pizzas, it will have to raise the price of electricity to pay for them; thus, applying a no-cross-subsidy test would prevent the utility from implementing an obviously cost-effective pizza programme.

When the utility service involves pizzas instead of light bulbs, most people will immediately see the logical flaw in this example: **Why is it assumed that the utility must GIVE AWAY pizzas??** If pizza advocates are correct that a $5 utility pizza is as good as the competing $8 pizza, the utility can easily sell pizzas for enough to prevent electricity prices from increasing; indeed, if the utility gives pizza consumers a $1 benefit by selling pizzas for $7, gives shareholders $1 as a reward and incentive for thinking of such a good idea, and uses the remaining $1 pizza profit to lower electricity prices, there are **no losers**!! If the utility cannot get customers to pay enough for its pizzas to produce such a result, then the claims about cost or quality must be wrong and utility pizzas are not cost-effective after all.

Of course, **some** utility pizzas may be cost-effective for **some** consumers even if the utility does not charge for them. But when consumers queue up for free pizzas, the utility and its regulators will have no effective way to decide which consumers really like pizza enough to deserve it, much less which should have the expensive special-with-anchovies rather than the standard cheese-and-tomato. The pizza experts will claim to know, of course, and will be quite willing to testify on the matter, telling the utility and its regulators how ratepayers' money should be spent; indeed, when hamburger-loving-consumers complain about paying for pizzas through higher electricity prices, the experts will argue for raising electricity prices even further to pay for free hamburgers so that everyone will have a give-away programme to defend. Happily, at least where pizzas rather than light bulbs are concerned, it is generally recognised that there are better ways to accomplish truly cost-effective results.

Substitute 'high-efficiency lights' for 'pizza' in this example and none of the logic changes. A utility can always give away or under price a costly service that allegedly 'reduces market imperfections;' if it does so, it will have to raise the price of some other service to make up its loss, whether kWh sales are higher or lower, and even if some of the under priced service manages to find its way into cost-effective uses. But this arithmetic tautology hardly invalidates the basic economic and business principle that any utility service that really 'reduces market imperfections' enough to lower end-use costs **can** be sold at a price that does not require such a cross-subsidy and **should** be sold at such a price in order to limit waste, fraud and abuse. And this is strictly an efficiency principle; the fact that it is 'fair' for those who benefit from a service to pay (at least) its incremental costs is merely a bonus.

Why is it taken for granted that costly and valuable demand-side services should be given away, or at least cross-subsidised by higher electricity prices? Why do so few people spot or question the critical pricing assumption implicit in literally every example 'proving' that cross-subsidies are necessary for cost-effective conservation, when almost everybody would recognise and reject the same assumption if it were applied to utility pizzas? This is not the place to speculate on the history, reasons or motives that have created this peculiar intellectual blind spot. But there can be no doubt that it is a blind spot or that it affects many analysts and regulators or that, like all blind spots, it is potentially very dangerous.

Levelling the tilted playing field

The assertion that demand-side programmes should be exempt from the well-accepted no-cross-subsidy test is really quite extraordinary. It is neither more nor less than the assertion that social resources should be allocated to demand-side programmes on the basis of hypothetical claims by interested parties who expect that, if they can just convince regulators that their favourite programmes 'reduce market imperfections' enough to be 'cost-effective,' costly services will be given away so that their claims about cost-effectiveness will never have to face a valid market test. This is a strange definition of a level playing field; there must a better one.

If conservation measures are really as incredibly cost-effective as advocates claim, they can easily be implemented with programmes that recover enough of their costs from participants to meet a no-cross-subsidy test. Consumers have had no trouble finding insulation, caulking and plastic storm windows in their local home centres, have found room on their credit cards to finance them, and have spent many a weekend installing them. These same consumers should be able to find, finance and screw in the $20 light bulb that, according to advocates, will reduce electricity bills by $2 per month for ten years with absolutely no other cost to the consumer.

Even if utility purchasing, installation, endorsement and/or finance are necessary to 'reduce market imperfections' enough to make this miracle light bulb cost-effective, there is still no justification for a cross-subsidy. The utility can include in its monthly bills information describing the $2 per month saving and inviting consumers to return a pre-stamped postcard accepting the entire bundled service for a $1 monthly surcharge - a price that will give the utility a nice profit to divide between higher shareholder dividends and lower electricity prices. Customers who will gain a net of $1 per month for simply mailing the postcard will do so in droves; even tenants and landlords suffering from the 'split incentives' so often cited to justify conservation subsidies should be able to find some way to divvy up the windfall. Consumers will decline the offer only when they know something about their specific situations, habits and tastes that makes them doubt they will benefit even by $1, i.e., when the light is not, in fact, cost-effective.

If advocates find it too onerous to let consumers participate in the process even to this extent, they can instead demonstrate to the regulators that a class of customers will gain $1 per month from the service, even after paying its incremental costs through the $1 surcharge on their class-specific bills. Regulators will jump at this chance to do ratepayers such a favour, knowing that any consumers who manage to figure out what has happened will thank them for the $1 per month windfall. The utility can then deliver the service, add the surcharge to the bills of the lucky beneficiaries, increase dividends and decrease rates.

Significantly, this picture of consumers, shareholders and regulators happily dividing up a free lunch is far from what is actually observed in the market and in regulatory proceedings, suggesting that the realities of conservation costs and benefits are rather different than the claims. Conservation services alleged to be incredibly cost-effective go unsold, often even when heavily subsidised; so demand-side advocates, in the name of maintaining a level playing field, advocate larger subsidies. Even the simple device of recovering easily-identified costs from easily-identified beneficiaries is opposed by demand-side advocates, presumably because consumers will not want the benefits if they must pay the (allegedly much-lower) costs.

Of course, consumers may decline to buy the latest conservation gadget or may object to the utility installing, financing and charging them for it simply out of ignorance or inertia. But it is also conceivable that consumers know something about their individual preferences and opportunities that the experts have not fully considered. Perhaps consumers do not want the high-efficiency light because of its colour balance or flicker, or because it looks silly in their fixtures, or because they do not use any single fixture enough to justify the $20 investment, or just because they do not want to be bothered for a potential and uncertain saving of a few dollars a year. Perhaps the landlord/tenant team knows that the commercial space will be remodelled soon, or that conventional lighting is a cheap way to keep clients and staff happy; anybody who has fought the automated lighting systems in some modern office buildings knows that conservation has costs the designers do not fully consider.

Even if the experts are, in some technical sense in some specific case, right about how other people should spend their money, there is some merit in the old-fashioned idea that those who will ultimately pay for something should have the incentive and the opportunity to compare the costs to the benefits and decide for themselves. Before having a claim on consumers' money, a utility's demand-side business should go through much the same process its supply-side business or its pizza business must go through before having a claim on consumers' money: determine what consumers want; develop services that meet these needs; find ways to deliver and finance these services without raising other prices; and convince consumers, individually or through their regulatory representatives, that the new service is worth having **and paying for**.

Of course, where electricity is under priced relative to its marginal cost, the competition between electricity and conservation is distorted. If it is impossible to redesign rates to reduce this market imperfection, there is a theoretical case for offering price-correcting subsidies that give consumers the right incentives to conserve - provided the inefficiencies inherent in any such subsidy scheme do not swamp any theoretical benefits. But even here, the no-cross-subsidy test applies: the highest conservation subsidy that can be paid without providing **too much** incentive for conservation is also the highest subsidy that can be paid **without increasing prices** above what they would otherwise be - the difference between marginal cost and price.

Today, the more common market imperfection due to mispriced electricity is a price that exceeds marginal supply costs, tilting the playing field **in favour of** conservation. In such cases a utility should reduce electricity prices selectively if necessary to avoid loss of profitable sales and should price any conservation services to earn the same margins earned on electricity sales. This 'second-best' solution also meets the no-cross-subsidy test, since cutting prices selectively if necessary to prevent uneconomic conservation will result in lower average prices than letting profitable sales go without a fight.

In summary, requiring a utility demand-side business to cover enough of its costs to keep electricity prices from increasing above what they otherwise would be, is perfectly consistent with truly cost-effective conservation, even when electricity is mispriced. To a first approximation, this means running a demand-side programme as an independent business or at least as a separate profit centre; any subsidies truly justified by electricity mispricing will generally be small and narrowly focused, and should be available to any supplier of demand-side services, not just the utility's own demand-side business. This is the only administratively practical way to maintain a truly level playing field.

Buying 'saved kWh' and demand-side bidding

The current interests in energy conservation and competitive purchasing of generation have led to proposals that utilities buy 'saved kWh' the same way they buy produced kWh; the 'megawatt auction' or all-source bidding process is the clearest expression of this idea. Of course, if such purchasing is done 'right' it will get the 'right' result; but this is a tautology that begs the real question of whether there is any right way to buy saved kWh that is worth the trouble or the risk that it will be done very wrong.

In the simplest and most distorted approach to buying saved kWh, a utility would invite offers of saved kWh alongside offers of produced kWh, selecting and paying for those kWh with the lowest bid prices per kWh saved or produced. This form of demand-side bidding involves an elementary logical flaw: buying a kWh is fundamentally different in economic and business terms from not-selling a kWh. When a utility buys a produced kWh for, say, four cents, the seller is compensated for the cost of producing the kWh and the utility gets a kWh that it sells on to a willing buyer for something near four cents; it is a simple exchange of value-for-value. But when the utility pays four cents for a saved kWh, the seller is paid for doing nothing except not buying a kWh for a fair price; four cents in value flows from the utility to the saver, but little or no value flows the other way.

Of course, a monopoly utility can, if its regulators allow or even encourage such nonsense, pay four cents for nothing and then recover its loss from its captive customers who have no say in the transaction and get no benefits from it. But this use of monopoly power to tax consumers hardly makes the value-for-nothing payment for a saved kWh equivalent in economic or business terms to the value-for-value payment for a produced kWh. Even if the utility is losing money on incremental sales, the value to the utility of not-selling a kWh is only what it is losing on marginal sales, which (unless the utility is giving away kWh) is nowhere near the full value of the kWh being offered by a supply-side bidder.

In effect, a utility always has a standing offer on the table to 'buy back' kWh at the retail price of, e.g., five cents: a consumer is always free to take one less kWh and keep five cents. If the consumer chooses not to sell back some kWh at the going price of five cents, it must be because it would cost at least five cents to get the same end-use service some other way. If the utility offers four cents for saved kWh in addition to the five cent standing offer, the total payment for saved kWh becomes nine cents, which will induce consumers to incur costs up to that level to save kWh. The utility will buy more saved kWh if it offers nine cents rather than five; but that does not make it cost-effective for it to do so.

Charles Cicchetti and William Hogan of Harvard's Energy and Environmental Policy Centre have developed a clever approach to demand-side bidding, based on the view that a utility provides end-use services rather than just commodity electricity (**Public Utilities Fortnightly**, June 8, 1989). As a provider of end-use services, a utility should charge for the service provided, not just for the kWh bundled in the service. Since a utility payment for a saved kWh allows the consumer to obtain the same end-use service with fewer kWh, it is only fair and efficient that the consumer should continue paying the same amount to the utility, i.e., should pay not only for the kWh it continues to consume, but also for the kWh it has been able to save as a result of the utility's payment.

Looked at yet another way, the only way a non-generating consumer can obtain any kWh 'resources' to sell to the utility is to buy more kWh than it consumes. Thus, the total kWh the saving consumer must pay the utility for is the sum of the kWh it consumes and the saved kWh it is paid for selling back.

So there is a 'right' way to include conservation options in an all-source bidding process: the utility selects the winning bidders on the basis of the bid prices, pays all winners the market-clearing price, but adds the saved kWh onto the demand-side bidders' meter readings and charges the retail price (less distribution-related costs) for them. Of course, if the retail price is approximately the same as the marginal supply cost (as it should be), winning a demand-side bid is not worth much because the saving consumer will pay as much to a kWh 'resource' as it obtains from selling it back. But this is just the appropriate reflection of the fact that consumers already have a conservation incentive equal to the retail price, and hence this amount should be subtracted from any payment made for conserved kWh in order to give the right price signals.

Demand-side bidding **a la** Cicchetti and Hogan could help correct some pricing distortions and finance conservation investments; the utility would make an up-front payment to a consumer who invests in conservation measures, with the loan/investment paid off through the monthly additions to the utility bill. This is a perfectly logical and economically correct approach to buying saved kWh; the fact that it is so different from the megawatt auction and 'qualified conservation provider' ideas being advocated in Congress and elsewhere merely demonstrates the economic flaws in these ideas.

The level playing field for environmental benefits

Finally, there are the environmental benefits of energy conservation - more accurately, the environmental costs of energy production. There is no doubt that energy production imposes costs on society through its effects on the environment, or that these costs should be taken into account in comparing generation and conservation alternatives. Nor is there any doubt that conservation on a large scale can be a very cost-effective response to higher energy costs; the experience of the last twenty years has demonstrated conclusively that energy demand will respond dramatically to price signals. The real issues concern **how best** to take environmental costs into account and, in particular, whether direct conservation **subsidies** are a cost-effective environmental policy compared to the supply-side and demand-side alternatives.

With insignificant exceptions, the **consumption** of electricity is environmentally benign; it is the **production** of electricity that damages the environment, and then to widely varying degrees, depending on how the electricity is produced. Burning strip-mined high-sulphur coal in an inefficient, uncontrolled power plant causes far more environmental damage per kWh of electricity produced than does burning natural gas in a high-efficiency combined cycle plant with nitrogen oxide controls. Subsidising the conservation of kWh independent of how the kWh are produced will be far less beneficial for the environment than incurring the same cost to conserve selectively those kWh produced in the most damaging ways.

In general, the way to obtain the most environmental benefit for a given total cost is to shift consumption from relatively 'dirty' to relatively 'clean' kWh, and then to conserve even the cleaner kWh to the extent justified by their higher cost and remaining environmental damages. The best way to accomplish this is to be sure that each production activity pays its full social costs, including the costs of appropriate environmental controls and any remaining environmental effects. This will encourage producers to seek cost-effective ways to reduce their environmental impacts and to pass through in consumer prices the costs of both environmental controls and remaining environmental damages, encouraging consumers to conserve energy with the appropriate emphasis on those sources that are most difficult to control.

Supply-side pollution taxes and marketable discharge permits are conceptually the best way to internalise environmental costs, although reasonable results can be obtained with other supply-side policies. However, demand-side subsidies for conservation attack the problem from precisely the wrong end; such subsidies provide no price signals or incentives for the decisions that directly affect the environment - decisions about fuels, technologies and locations used to produce the kWh. Paying a consumer four cents to induce conservation of a random or 'average' kWh, if paying one cent to an equipment vendor would convert a dirty kWh into a clean kWh, is bad economics and bad environmental policy; it accomplishes less than it should and reduces society's willingness and ability to solve environmental problems.

The inefficiency of attacking supply-side environmental problems with demand-side programmes can be illustrated with the control of sulphur oxide emissions from power plants. An emission charge of twenty-five cents per pound of sulphur dioxide would almost surely reduce sulphur dioxide emissions by fifty percent or more. The resulting higher consumer prices would also encourage conservation; but the total effect on consumer prices of the control costs plus the charge on remaining emissions would be less than a cent per kWh, even for a utility that was originally relatively 'dirty.' If this price increase reduced electricity sales for this dirty utility by ten-to-twenty percent, it would reduce emissions an additional five-to-ten percent of the original level, compared to the fifty percent reduction likely to be accomplished on the supply side at the same per-unit cost.

Put another way, twenty-five cents spent on electricity conservation for which a utility must pay four cents/kWh would save about six kWh. If it were possible to assure that only very dirty kWh were conserved, this could reduce sulphur dioxide emissions by 0.2-0.5 pounds; if only 'average,' post-supply-side-control kWh are conserved, the reduction in sulphur emissions could be an order of magnitude less. The same twenty-five cents spent on scrubbing of the remaining dirty sources could remove several pounds of sulphur. Thus, demand-side measures to reduce sulphur emissions could easily be ten-to-twenty times less cost-effective than supply-side policies that attack sulphur emissions directly.

Even for reducing carbon dioxide emissions, where conservation is often regarded as the only effective solution, subsidising conservation directly is likely to be an inefficient policy. Conserving one kWh will save roughly half a pound of carbon if the saved kWh is produced from typical coal, half as much if the kWh is produced from natural gas in a combined cycle plant, none if the kWh is produced from nuclear or hydropower. If a subsidy of four cents per kWh is necessary to induce conservation of one kWh, the implied control cost is about $160 per ton of carbon if only coal-generated kWh are conserved; if an 'average' kWh is conserved, the implied carbon control cost could be well above $200 per ton of carbon.

A $200 per ton carbon tax/rebate would increase power plant coal prices by a factor of three to five and would stimulate a dramatic and costly supply-side response: some combination of massive shifts to less carbon-intensive fuels, application of highly-efficient fossil fuel burning technologies, removal of carbon dioxide from stacks with disposal in the deep oceans or underground, tree planting on a global scale, etc. Whether society should be or will be willing to pay costs of this magnitude to control carbon dioxide is a question well beyond the scope of this analysis; after all, even adjusting to a warmer climate is an alternative with finite costs. But if society is not willing to bear carbon control costs of this magnitude on the supply side, there is no reason it should bear them on the demand side either.

If society is willing to pay carbon control costs at the level suggested by a $200 per ton carbon tax, to what extent is electricity conservation part of the most cost-effective response? There has been little analysis in this area, but some illustrative calculations are instructive. A $200 per ton carbon tax would increase average electricity prices by (very) roughly four cents per kWh if there were no supply-side response, and less to the extent there are supply-side control measures costing less than $200 per ton; suppose the price increase is 2-3 cents per kWh, amounting to some 30-50 percent of average retail prices. Such a price increase might reduce electricity consumption 20-30 percent, reducing carbon emissions by this fraction of the emissions remaining after taking account of the supply-side response; i.e., if the supply-side response reduced emissions per kWh by half, conservation would contribute another 10-15 percent. Overall, the supply-side response in this example would be 4-5 times as large as the demand-side response.

These are illustrative figures only, not intended to prejudge the question of the importance of conservation in an overall environmental control policy. If supply-side measures are much less cost-effective than assumed above, conservation will have to bear the brunt of any emission control society is willing to pay for. But it is likely that supply-side measures will dominate a cost-effective approach to environmental control even where carbon dioxide is concerned, and that focusing on demand-side subsidies may turn out to be very costly and ineffective.

The potential costs of environmental control are far too high for society to tolerate inefficient policies. The emphasis must be on supply-side policies that attack these problems at their sources and cause downstream price signals to reflect upstream environmental costs. These downstream price signals will make utility (and non-utility) demand-side businesses profitable where conservation is a cost-effective response. However, if such direct environmental policies result in less energy conservation than demand-side advocates would like to see, this is not an argument for additional conservation subsidies. Quite the contrary: it is evidence that forcing or bribing consumers to save energy is not a cost-effective approach to environmental improvement.

Conclusions

Utility demand-side programmes deserve to be given full and careful consideration as a potential way to give consumers better end-use energy services for their money, utilities an edge in an increasingly competitive world, and society a way to reduce the environmental costs of energy production. But in each of these areas, demand-side programmes offer no free lunches and have no inherent advantages over supply-side programmes. There are many ways for consumers to meet their needs, for utilities to prosper and for the environment to be protected; all options, whether inside or outside the utility and whether on the supply-side or the demand-side, should be considered on a truly level playing field that is not tilted and confused by misleading jargon and concepts.

If utility demand-side programmes make sense on economic and business grounds, they should be able to meet the standard economic and business tests applied to other utility and non-utility activities. If utility demand-side programmes make sense on environmental grounds, they should be able to demonstrate their cost-effectiveness relative to other, primarily supply-side, measures society is willing to undertake to control environmental effects.

If this all sounds simple and obvious, it is. There is nothing about utility demand-side programmes that cannot be understood by applying the same economic and business common-sense that works for utility supply-side programmes and for non-utility activities.

Reference

Cicchetti, C and Hogan, W (1989), Public Utilities Fortnightly, June 8

Discussion: Public policy towards new investment

Nick Woodward

The topic of 'Public Policy Towards New Investment' begs the question 'in what?' So our contributors have rightly defined it for themselves. Jim Skea has provided a careful mapping, with comparative examples, of the linkages between investment, technical change and environmental regulation, underlining the necessarily contingent nature of practice where market conditions and technology are changing, and where existing configurations constrain future choice, with a consequent need for flexibility and sensitivity in regulatory relationships and practice. Professor Yamaji's simulation suggests that subsidies to promote CO_2 reducing investments, in his case at least, are likely to prove more efficient and effective than a carbon tax. He too underlines the contingent nature of the Japanese context - a national consensus on the need for energy efficiency and for pollution reduction, and good government-industry relations which encourage informed and consensual decision-making in regulatory affairs. Larry Ruff has argued persuasively for a 'business-like' approach to energy saving which recognises the reality of consumer decision-making, and the unreality of 'populist' solutions from regulators and utilities which have political, but not economic logic.

What do these diverse papers have in common? First, a recognition that economic agents behave economically, reacting and adjusting to price/cost signals, to regulations, over time. This apparently obvious point constitutes a powerful argument against command-control/administrative interventions which tend to ignore systemic effects, even though they may demonstrate political pay-offs. Three points follow. First, the need for economic thinking (not a priori theory) - the working through and evaluation of possible consequences in particular contexts. Second, the critical importance (pace regulatory capture) of consultation and information exchange, to explore the feasibility, costs and consequences of regulatory initiatives. Third, the particular contextual nature of different countries' conditions suggests that there are no standard 'ideal' solutions, but there is likely to be great benefit from comparative analysis, which is likely to provide grounded specification of 'good' regulatory processes and instruments, with broad guidelines, much as has occurred in the case of privatisation programmes.

Two additional points about new investment are worth making. First, current pollution reducing regulations and technologies in energy focus on known pollutants (particularly oxides of sulphur and carbon). But it is conceivable that in reducing these through new or retrospective investment (or switching fuel source), a further set of pollutants may be generated which will also prove harmful, generating a new regime of regulation and investment. This possibility suggests a strategy of looking for flexible/adaptable technologies and open-minded evaluation of possible consequences: but there is currently no economic incentive to develop or invest in technologies which minimise all pollutants (if this is conceivable), given particular resource inputs. Second, technological innovation in economics tends to be treated as an exogenous process (like entrepreneurship). Analysis of the generation of new ideas and of technological development lies more in the domain of history, of psychology and sociology, and depends much on contingent, contextual conditions. These conditions may be more or less favourable, as in the upswings or downswings of Kondratieff cycles, or in cycles of industry growth and development.

My feeling is that, at the moment, the conditions for radically different technologies are relatively unfavourable as energy generation tends to be concentrated in production, and oligopolistic (if not monopolistic) in form, with barriers to entry relatively high and the technical and economic potential of alternative technologies evaluated against existing practice, itself the outcome of decades of development. This phenomenon of 'lock-in' may generate a degree of technical and evaluative conservatism, which would be lacking in markets with low entry barriers and dispersed competition. Existing technologies and practices are maintained (and improved incrementally) by those whose professional experience and expertise has been built on them. Ironically, nuclear power generation in the 1950's was imposed on an unwilling industry by political will: it now has an industry constituency. The development of internal combustion in the 1890's depended upon a fortuitous combination of circumstances which at a critical juncture favoured internal combustion over steam: and the standard gauge railway can be traced back to the gauge at George Stephenson's Killingworth colliery, though Brunel's later broad gauge offered significant advantages in safety and operation.

Judgements on technologies are made in the light of current knowledge and economics, and their development depends somewhat on circumstance. "I think there is a world market of five computers," said Thomas J. Watson, president of IBM, in 1948 - and he was right, in relation to their then cost and application. If one acknowledges this, then the energy industry, as currently constituted, may have a bias towards existing technology and energy sources, which may also influence evaluation of alternative technologies at an early stage of development. And what firm or industry would invest in research into technology which might threaten its market position? Presumably domestic generating sets based on renewable energy sources are conceivable, but currently not feasible, technically and economically - like computers in 1948. But whether research into radically different technologies should be left to industries, firms, universities, governments, or inventors, is a matter for contingent industrial policy.

PART 5

Problems of Multi-regulation

CHAPTER 12

Multiregulation of energy and the environment: The case of Germany

Wolfgang Pfaffenberger

Markets, institutions and the environment

Emissions from the use of energy are one of the important environmental problems of our present time with strong implications for human health, plant and animal life and the global climate.

Due to political pressure in many countries, regulation has been developed to cope with at least some of the problems. Because of the high costs of emission abatement, the economic and ecological efficiency of regulation is a difficult problem but policies seem to have been influenced much more by political efficiency in the sense of impressing the voter.

The following figure gives an overview of the toolbox of environmental policy and lists the instruments by their market (efficiency) orientation. Until now market oriented instruments have played a rather small role in environmental policy.

German environmental policy has a high reputation for its high standards. The paper will give an outline of the various instruments employed and the results achieved. Some of the results are impressive. The efficiency record, however, is not. Regulation has become more and more complex and we do not really know what the cost of this regulation is in relation to its benefit.

Figure 1

Energy and environment regulation

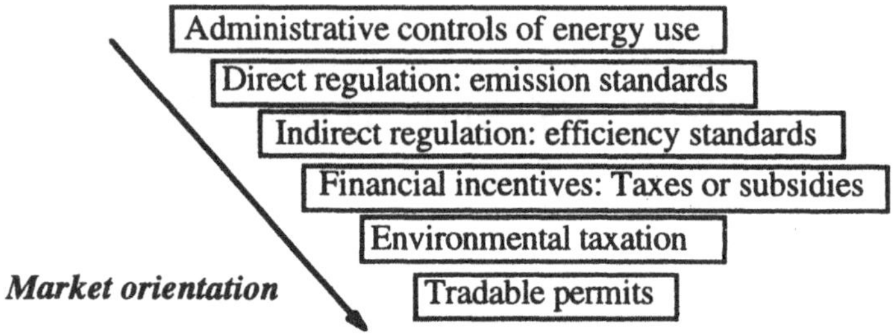

German economic policy has traditionally been in favour of market forces as a source of economic development. There is a strong contradiction between this general idea and the lack of market oriented institutions to cope with environmental problems. In the institutional framework of a market economy, environmental goods have to be taken care of by special agencies. But this does not mean that the cost of using environmental factors could not be internalised.

The alternative concept to present day regulation is a market approach. Instead of interfering with the action of individual agents in a detailed way, it would be possible to put a scarcity value on environmental factors by taxation or even better by restricting their quantities. Market forces would then take care of the rest. A lot of detailed regulations (and their regulators) then would be unnecessary because the price mechanism would help economic agents to find their own way within the restricted area.

Energy and the environment

The interactions between the energy system and the environment are very complex. When we think of emissions from energy use we usually think of the emissions contained in the smoke from burning different types of fuels. This is the typical type of emission that arises as a by-product of energy transformation or the end use of energy. But also, the production of primary energy has a number of harmful implications for environmental factors. These may be related to the soil, to ground water, use of landscape etc.

Figure 2 shows the flow of energy in the energy system and the connected emissions in the different levels (production of primary energy, transformation, and use). It is typical for the energy system that considerable amounts of the primary product are being used up in the process of transformation so that a large share of the emissions is not directly attributable to an energy service consumed. Given the scarcity of environmental factors from an economist's point of view, the problem is to find the optimum emission regarding the whole energy system. So it is the total cost of all energy uses and the total emission (in the figure: cumulated emission) that we are interested in. Economic agents on the different levels of the economy, however, consider only their own sphere. The same is probably true for regulators who concentrate on the specific factors that are quite different for primary energy production, transformation and use of energy. It is very unlikely that a multiregulation approach with specific regulations for all levels of the energy system will lead to anything near to an economic optimum.

On the other hand, it may be very practical for the policy maker to start at some easy-to-define points in the system which are deemed relevant for the emission problem and then develop a more and more complex system of regulation. This approach may be very successful in the beginning because a fast change can be achieved in a short time (as was the case in Japan and also in Germany). Later on, however, you may get lost in a complex system of regulation. The side effects of many of the different regulations may cancel each other out without the regulator even noticing it.

The institutional setup in Germany [1]

The main instruments used at present are of an administrative kind. In Germany there is a tradition of regulation in the energy sector that had its roots in the coal problem. Highly expensive German coal is protected from international competition by a system of subsidy that is paid for by the consumers of electricity.

In contrast to the highly regulated electricity industry, the market for oil products and natural gas is not regulated with respect to quantity or price.

Regulation of emissions from energy use came about as a result of the acid rain problem which was a strong political issue for some years. At that time policy makers concentrated on the electricity sector (which was regulated anyhow and has a high degree of public ownership). Other sectors followed later.

The oil crisis led to some indirect forms of regulation. The aim of these regulations was to increase energy efficiency but improved energy efficiency, as a side effect also leads to reduced (specific) emissions.

The most problematic sector nowadays is the transport sector. The large number of emitters (and voters), and the importance of the automobile sector for employment and exports, leads to a certain reluctance in applying regulation.

Figure 2

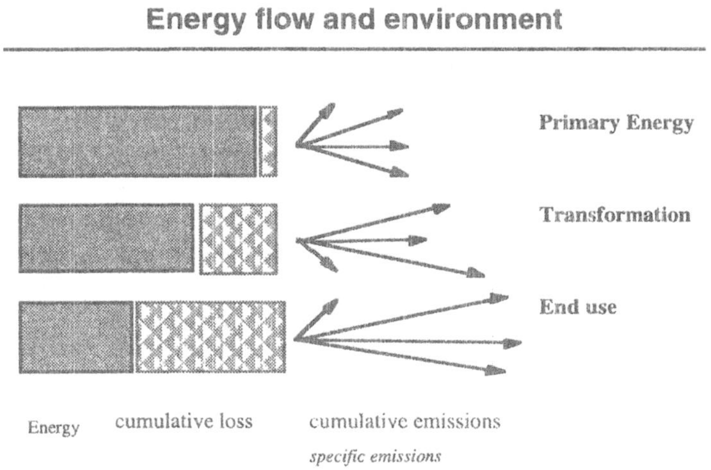

The German state has the form of a federation. The three government levels of federation, states and municipalities all play a certain role in the regulatory process. In the following, a short overview of the power of these three levels will be given. Some very special problems came about with the unification of Germany. These will be explained in the second part of this section. The third part will comment on the regulatory instruments used.

Germany as a federation

The division of powers in the German system of government has many implications for the issue of environmental regulation. Figure 3 gives an overview of the role of the different parts. The federal level is responsible for general legislation on environmental issues and also maintains the federal environmental office. But the latter is not really an administrative body, but rather an institution to promote environmental knowledge and environmental thinking in and outside government. The basic approach in the federal environmental protection law (Bundesimissionsschutzgesetz) is to set standards for certain environmental factors and set up an administrative procedure for implementing these standards.

Figure 3

Germany as a federation

Federal level
General Legislation, sometimes as framework, most important: Federal environmental protection law.

State level
Environmental administration and monitoring, financial incentives.

Local level
Right to supply energy. Income from local energy tax and profits from energy companies often used to subsidize public transport.

Implementation to a large extent is done by the state governments and their regional branches. The basic rule is that activities that fall into the scope of the law need to be permitted. The most important problem of implementation, therefore, is the administrative practice as regards these permissions.

Also the local level is strongly involved in energy matters. It is generally agreed that local governments have the right to decide on the supply and distribution of electricity, natural gas and district heating within their territory. Often local governments own utility companies to fulfil this duty: sometimes they have contracts with other utilities.

Local governments have very strong financial interests in the energy business because they can tax energy sold (particularly electricity) and usually use this income to subsidise public transport or other services. Declining sales of energy due to energy conservation are a nightmare for every local government treasurer because they would withdraw the financial basis from other local government services. Politically this is a very touchy issue.

Some people think that strengthening the power of local government would automatically lead to environmental improvements due to the decentralisation effect. Many of the states have created so called energy agencies.[2] These are meant to promote decentralisation of energy production and distribution by supporting the planning and implementation of Combined Heat and Power (CHP), energy conservation etc. on the local level.

Special problems of unification

Energy regulation in Germany has developed over the last two decades as a continuous process of increased complexity of regulation. Targets were changed and adjusted according to development. The new German states have a completely different history. The main line of East German energy policy before unification was to reach the highest possible level of self-sufficiency using domestic brown coal to the highest possible degree. The Russian/East European system has been described as 'the economy of shortage'.[3]

The typical shortage of investment funds within the system led to under investment with the result of high emissions from the energy sector. Energy efficiency remained at a low level and the number of installations for emission abatement remained small.

According to the unification contract, West German law will be applied from 1996 on. This will have very strong implications for the transformation sector: most equipment will either have to be replaced or switched to natural gas instead of coal (in most cases temporarily). Some of the problems will be solved automatically by adjusting the transformation sector because the degree of district heating in East German Cities is rather high. On the other hand about half of the stock of dwellings is still equipped with traditional stove heating on the basis of coal. The gradual switch to central heating will automatically lead to the use of cleaner fuels (heating oil or natural gas).

In the transport sector, reaching higher standards in specific emissions is not a big problem because the rate of car ownership was relatively small in the former GDR, and the many new cars currently purchased will fulfil the standards. The shift of transport from rail to road and the increased transport activity will, however, raise the level of emissions.

An important aspect for future development could result from the decision to restrict coal protection policy to the old Länder. In the new Länder, theoretically this could lead to a greater priority given to natural gas in the production of electricity. On the other hand, in the regions with high production of brown coal, there is a strong economic dependence on continued exploitation of this resource. This might lead to a new kind of coal policy for the new Länder.

Some more details on the differences in emission between East and West will be given later when we have a look at the figures.

Basic outline of environmental regulation

Basically environmental regulation in Germany follows the philosophy of administrative control by permission and standards. The federal environmental protection law [4] is based on the idea that the best way to protect people from emissions or risks is to control the construction and use of equipment that can generate such emissions or carry certain risks.

For certain types of equipment (large scale, higher risk), a special permission is necessary to construct such equipment. Once the equipment has been established, monitoring takes place to see whether the standards for emissions are being fulfilled.

Smaller and less risky equipment can be constructed without permission but there is a general rule that environmental emissions that can be avoided according to the state of the art should be avoided. The monitoring of such kind of equipment is less strict. The law gives room for detailed administrative controls also for this type of equipment if deemed necessary. Quite a number of such detailed rules exist now and will be commented on below.

The federal environmental protection law that was created more than 20 years ago in the meantime has become a complex system of regulation. The basic idea is that to promote certain environmental standards you have to influence investment. The influence of environmental policy is particularly strong when new equipment is being constructed.

From the point of view of the regulator this makes sense. From the point of view of the regulated this is less clear. There is no incentive to do anything except what is required by the standard. As the law refers to the state of the art, the incentive could be **not** to engage in technical progress regarding the reduction of environmental emissions because this would lead to higher standards in the future. In addition this type of regulation promotes end-of-pipe technologies but does not promote different or new processes of production that avoid certain types of emission.

In the following section I will give an overview of all regulations that refer to energy and the environment.

Regulation of energy and environment: A survey

Most of the regulation reported in the following is based on the federal environmental protection law. In addition there are some other institutions that are relevant for our subject like the federal law for the protection of nature (Bundesnaturschutzgesetz) and others.

For new plants an environmental impact analysis has to be carried out now that has to look at the local implications of the plant in addition to the general standards.

Primary energy production

The production of coal is the most dominant production of primary energy in Germany. Both the production of hard coal in deep coal mines and the production of lignite particularly in the East from the surface have strong implications for the local and regional environment.

Coal mining is regulated by a special law and nowadays the manifold implications of coal mining are regulated with the help of the environmental impact analysis. On the basis of such an analysis, administrations can require special compensation in exchange for the right to exploit coal resources. This rule is based on the natural protection law (Bundesnaturschutzgesetz). This kind of compensation may reach from small scale investment to very large measures of recultivation with very high cost and can take many forms between restitution on the site and spending money on natural protection at some other site.

Generally authorities can make use of the fact that for a certain activity, a permission is required to ask for side payments for the benefit of nature. The environmentalist movement often helps indirectly to increase the price authorities can charge for the use of environmental goods. There are signs that this type of bargaining will play an increasing role wherever problems of location have to be solved although this was probably not the original idea that authors of these laws had in mind.

Special equipment rules

The following equipment needs a special permission according to the federal environmental protection law:

- All heat and electricity generators ≥ 50 MW
- Most electricity or heat generators from 1 to 50 MW (depending on type of fuel)
- Cooling equipment for power stations using more than 10.000 m3/h

- Wind converters $\geq$ 300 kW
- Transformer stations $\geq$ 220 kV
- Coal processing equipment of different types
- Oil refineries

For this type of equipment, construction and use require a special permission. Before such a permission is granted, the authorities have to check the environmental effects of this type of equipment. For a number of important emissions standards have been set. This will be explained below.

For water power the special rules of the water protection law hold.[5] The same applies to storage of energy resources especially oil and oil products.[6]

Risk regulation

A similar kind of approach has been chosen to cope with environmental risks. Equipment that can be used only on the basis of special permission can also be subjected to the special risk regulation. This regulation can require particular measures of precaution to minimise technical risks of the equipment. In addition a special monitoring of emergencies takes place. In the energy sector refineries and certain types of coal processing plants are subjected to this kind of risk regulation.

Product standards

For a number of products special product standards have been defined. These refer to:

- The sulphur content of heating oil and Diesel oil (0.2 % of weight).[7]
- The lead content of gasoline (0.0013 g/l).[8]
- No bromine and chloride in gasoline/Diesel oil.[9]
- Coal pellets may only contain certain additives. Coal used in residential heating has to have a limited content of sulphur (1 % of substance).[10]

Emission standards

Standards have been defined for a number of emissions. Some of them apply to equipment that is subject to a special permission, others also apply to other equipment. The monitoring is different depending on the type of equipment.

Dust

Regulation differs between large equipment (≥50 MW) and smaller equipment. The amount of particles allowed also differs according to the type of fuel (coal, oil, gas) and the type of equipment (gas turbines are treated differently).[11] Special references are made to old equipment for which different standards are valid.

The regulation for dust is relatively strict so that dust emissions from energy equipment are not relevant today in relation to other sources in the 'Old Länder'.

In the transport sector special standards have been set for diesel engines (See below under NO_x).

Sulphur dioxide

SO_2 was greatly debated in Germany when the acid rain problems became apparent. SO_2 emissions from heating oil are controlled by the product standards mentioned above. The most important source of emission is the burning of coal. Regulation is similar to the one for dust. The difference is that special equipment is necessary to reach the standards prescribed for large scale equipment. Coal fired power stations above a certain size nowadays usually use end of pipe technologies to clean the flue gas. Smaller stations sometimes use flue bed technologies to reduce the sulphur in the process of burning. As a by-product, secondary materials have to be taken care of. The most important by-product is gypsum. This is now being recycled into the building sector.

For old equipment there was a choice between retrofitting or closing after a certain deadline.

Many modern plants can over fulfil the standard easily. Local authorities may require higher standards due to local emission considerations. If the abatement equipment breaks down the power station can continue generation for up to 72 hours: the maximum number of hours without cleaning is 240 hours per year.

Nitrogen oxide

The situation is very similar with that for SO_2. Until now there have been no standards for small scale burning equipment but they are expected in a new amendment.

The most important emitter of NO_x is the transport sector. Emissions from this sector are now regulated in the following way: standards have been defined for emissions of NOx, CO and VOC for different types of cars. For diesel engines standards have been set for the dust emissions. Cars need to be licensed anyway, so the environmental standards were incorporated into the general conditions of licence.[12] Cars are classified by the level of emission. Low emission cars are eligible for a reduced tax rate. Monitoring takes place every year.

VOC

VOC from energy use to a large extent results from incomplete burning of fuel in the transport sector. Another source is the evaporation of components from storage, transport and distribution of fuels. The first type of emission is taken care of by the condition of licence for cars (see under 4.5.3), for the second type two directives have been issued in 1992. They regulate measures for avoiding the emission of VOC when pumping fuel from storage to means of transport or from gasoline station to cars (special nozzle to withdraw evaporated fuel).[13]

Carbon monoxide

For carbon monoxide standards have been defined for small and large scale burning equipment and for cars.

Carbon dioxide

No regulation for the emission of carbon dioxide exists. The Federal Government decided in 1987 to reduce the emissions of carbon dioxide by 25 % to the year 2005. This declaration is still valid but it is unrealistic to expect that this reduction can be achieved within the next 12 years.

The usual standard approach does not work with CO_2. Typical instruments are energy substitution or energy conservation. Due to the high share of coal in electricity production and the high level of per capita consumption of energy in Germany the per capita emission is relatively high (for European standards). There is a strong conflict between CO_2 reduction goals and coal policy. I will comment on that later on.

Efficiency standards

Environmental impacts of energy production and use can be reduced by reducing the consumption of energy itself. The regulation reported here was not originally designed for environmental purposes but is a typical product of the high oil prices in the seventies with the aim of reducing the consumption of heating oil or other fuels. In the meantime this regulation is being used for environmental purposes.

Heat as a by-product

Many technical processes produce heat as a by-product which is not or cannot be used directly. The federal environmental protection law as a general rule has formulated that heat as a by-product should either be used within the same process of production or made available to third parties as far as this is technically feasible and can be expected according to type and location of plant and is not in conflict with other aims of the law.[14] The federal government has the right to issue directives on the application of this principle.

Until now one directive has been issued concerning the burning of waste. If heat cannot be used directly or used by third parties, it must be used for producing electricity if the capacity available is at least 500 kW.[15]

A more general directive is being prepared. Industrial and commercial users will be required to develop a concept for the recycling of heat within economically defined limits.[16]

Energy efficiency in buildings

For new buildings energy efficiency standards have been set on the basis of a special law created in 1976.[17] The efficiency standards relate to new buildings. For existing buildings it has been prescribed, however, that energy consumption in apartment buildings has to be measured apartment by apartment and energy bills have to be based on consumption by individual users. This measure has been carried out in the meantime and it has been effective in the sense that tenants of apartment buildings now have an economic interest in energy conserving behaviour.

The efficiency standards for buildings are currently under review. There is a strong conflict between the environmentalists who want higher efficiency standards and the bricks industry which refuses to change their products correspondingly.

Efficiency of heaters

Residential and larger heaters have to fulfil certain efficiency standards that differ by the age of the equipment. The monitoring is done once a year by the local chimney-sweepers. In October 1993 the efficiency standards for older heaters will be raised. It is expected that this will lead to a modernisation boom.

Promotion of substitutes

The production of renewable energies (although not necessarily the production of equipment to produce renewable energy) is almost free of emissions of the kind associated with fossil fuels. To promote renewable energy, a law was passed in 1990 that sets the minimum price for electricity from renewable sources at 90 % of the average value of all electricity sales.[18] In addition, wind energy and to some extent also photovoltaic generation of electricity is subsidised by the federation and the states.

Although the law required that electric utilities have to buy electricity from renewable sources at the legal minimum price it did not regulate the cost of transmission lines between the site of production and the electricity grid. Wind energy right now in Germany is booming and there are a lot of disputes between utilities and investors about the cost of transmission lines.

There is a strong conflict between energy regulation that wants to promote emission-free renewable wind energy and other concerns like the protection of landscape and nature. In the Northwest of Germany several hundred MW of investments in wind energy are expected for the next years. The state government has recently issued a directive for the regional authorities which aims to reconcile both regulative interests.[19]

Environmental effects of protective measures

As was mentioned above, the production of (very expensive) hard coal in Germany is protected by government intervention and a system of subsidy which is paid for by electricity consumers. To protect the role of coal in the electricity industry, it is not permitted to run power stations on the basis of natural gas or oil for medium or baseload purposes. So one of the very simple answers to environmental emission, the substitution of fuels, in this very important sector is not available.

The future of coal regulation is unclear. The present regulation will end in 1995, but the government has promised miners to continue a similar kind of protection with a somewhat reduced level of coal production.

The protection of hard coal is restricted to the 'Old Länder', but does not include the 'New Länder' where a boom of gas fired combined heat and power producing units is expected in the next few years.

Of course we do not know what the fuel mix in the electricity sector would be today if the coal market had been liberalised in the past. Because electricity produced on the basis of world market coal is the cheapest option available, it is quite likely that the share of electricity would not have been lower in a liberalised framework. Therefore environmental policy to cope with emissions from coal fired power stations would have been necessary anyway. In the future this could be different, however, for coal has the highest carbon content of all fossil fuels. There is a strong contradiction between the political will to reduce CO_2 emissions on the one hand and promoting the use of coal on the other hand. The shadow price of restricting the CO_2 emissions from the production of electricity is extremely high when the present coal policy is continued and nuclear power cannot be used extensively.[20]

Figure 4

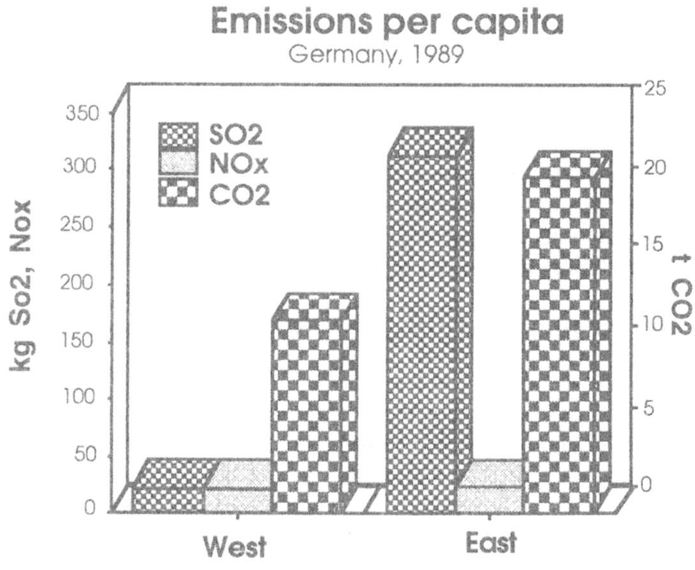

The development of emissions and imission (discharge)

Emissions

The following relates to the figures in the annex. It is necessary to look at the 'Old' and the 'New Länder' separately because of the different policies followed in the past. In the 'Old Länder' the most remarkable success was in the reduction of SO_2. Due to the strong reduction in the power industry, emissions are now at a level of about one quarter of 1970.

Figure 5

SO_2 - Emissions

Tonnes per square kilometre without transport

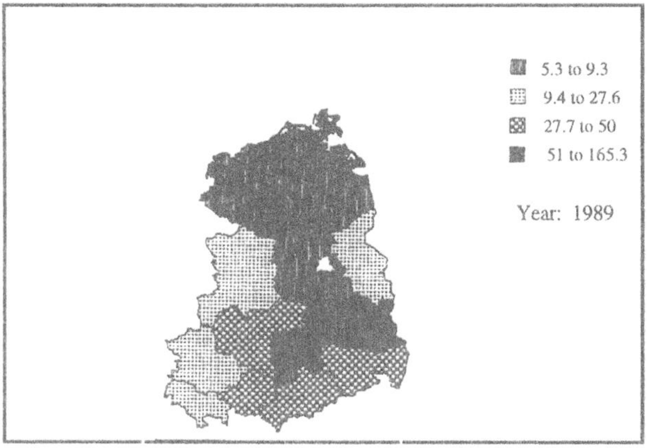

Source: Statistical Office of GDR

The emission of NOx has almost been constant with the share of the transport sector increasing continuously. Environmental policy was successful in reducing emissions from the power industry but in the transport sector, the reduction of specific emissions was compensated by increased transport activity. The transport sector also is responsible for most of the emissions of volatile organic components. The overall emission here is also almost stable.

Figure 4 shows the per capita emissions in Germany in 1989 - the year before unification. The picture shows the enormous difference in the emission of SO_2 between the two parts of the country, the per capita emission being about 15 times higher in the East. Due to regional concentration of emissions, some regions had to bear a very high load of sulphur emissions as can be seen from Figure 5.

The per capita emission of NO_x (without the transport sector) is about the same and the emission of CO_2 is much higher again in the East because of the high share of coal in the consumption of energy.

As was mentioned above, the power industry will have to be reconstructed in Eastern Germany in order to fulfil the standards by the year 1996. This will be partly achieved by substituting fuels (at least for some time) and partly by rebuilding power stations. Eastern Germany has a high share of district heating in the larger cities but in these areas the rate of CHP was relatively small and can be increased. It is expected that a number of modern combined cycle plants on the basis of natural gas will be implemented. We can expect a dramatic reduction in the emission of SO_2 and CO_2 in the next few years.

Imission (discharge)

Imission depends on emission within the country and the balance of imports and exports between countries. Such a balance requires model calculations. In such model calculations, Germany (in the old borders) appeared as a net importer of SO_2. The inclusion of the 'New Länder' has made Germany a net exporter. (see fig. 14 and 15). With regard to NO_x Germany has been a net exporter in both parts and remains a net exporter.

The average freight of SO_2 is about two t/km^2 in the old Länder and was about four times higher in the new Länder. The freight of NO_x is about 1 t/km^2 in both parts of Germany (see figure 16). It would too long to go into details about the regional distribution, but generally the level of emission is correlated with the density of population and industrial activity.

Critical evaluation

Cost benefit aspects

Widely varying estimates of the social cost of energy use have been produced. There are numerous methodological problems connected with each approach. For emissions with local and regional effects, the social cost is not independent from the location of the emitter, for emissions with global effects it is hardly possible to include all the interrelated effects that could arise from changes in the ecosystem.

It is clear that emissions result in damage. If emission is reduced, the estimated damage will be reduced too (although we do not know the shape of the damage function). If the cost of reduction of emissions is smaller than the reduction in damage environmental policy makes sense economically. In a recent study this was shown to be true particularly for SO_2 policy in Germany.[21] The benefit was found to be twice as large as the cost in this case.

Efficiency aspects

However, whatever the benefits of environmental policy, the question has to be asked whether these benefits have been achieved at the lowest possible cost. Theoretically it is clear that the administrative type of policy followed in Germany cannot be cost efficient, but there is no empirical demonstration of such theoretical insights. There have been some studies that showed the relative inefficiency of the standard approach.[22] These cannot be generalised. In this case efficiency has to be seen as dynamic efficiency i.e. the capability of the system to adopt to environmental restrictions. Neo-classical analysis does not help us very much in this respect.

From an institutionalist point of view, there is much to be said for a very distinct line to be drawn between the regulated and the regulators. It is the task of regulators to define environmental restrictions on the best available knowledge about ecological feedback. But it cannot be their task to regulate the activities of the regulated in detail.

Conclusion

The administrative approach to environmental policy that was almost perfected in Germany during the last twenty years has reached its limits. Success was tremendous in the reduction of SO_2. It is a sphere where success can be expected: a small number of emitters, well-known standardised technology of abatement and economic agents with sufficient financial power to finance the necessary investment and economic monopoly which allows shifting of the cost to the consumer.

These conditions are very special. The opposite conditions are true for the transport sector which is responsible for most of the NO_x emissions. The administrative approach is completely unable to cope with CO_2 and other greenhouse gases. Reduction of CO_2 requires different kinds of substitution and cannot be achieved by adding end-of-pipe technologies to existing processes.

Environmental policy in Germany will have to learn new lessons in the future but this may take time, particularly due to the many economic and social burdens implied in the process of unification.

References

Fünfter Immissionschutzbericht *der deutschen Bundesregierung*, Deutscher Bundestag 12/4006, 15.12.92

Kornai, J:. Economics of shortage, Amsterdam., 1980

Lübbe-Wolff, G.: *Vollzugsprobleme der Umweltverwaltung,* in: Natur und Recht, 1993, 217 - 229

OECD: *Environmental Performance Reviews:* Germany, Paris 1993

Pfaffenberger, W. u.a.: *EG Binnenmarkt und räumliche Energieversorgung,* Idstein 1992

Pfaffenberger, W.: *Economic effects of* CO_2 *reduction in Germany, in: Kuckshinrichs, W. u.a.:* Workshop on the economics of the greenhouse effect, modelling strategies and impacts, Bad Zwischenahn 1993, p. 15-32

Umweltbundesamt: *Daten zur Umwelt 1990/91*, Berlin 1992

Welsch, H.: *Kosten der* SO_2*-Minderung unter alternativen umweltpolitischen Strategien:* Ergebnisse eines Simulationsmodells des westdeutschen Kraftwerkssektors, in: Zeitschrift für Energiewirtschaft, 1989, 51 - 59

Witte, H. u.a.: *Umweltschutzmaßnahmen und volkswirtschaftliche Rentabilität,* Umweltbundesamt Bericht 4/92, Berlin 1992

Annex: Laws

Gesetz über Naturschutz und Landschaftspflege (Bundesnaturschutzgesetz-BNatSchG) last change 12.02.1990

Gesetz zur Ordnung des Wasserhaushalts (Wasserhaushaltsgesetz-WHG) last change 12.02.1990

Gesetz zum Schutz vor schädlichen Umwelteinwirkungen durch Luftverunreinigungen, Geräusche, Erschütterungen und ähnliche Vorgänge (BundesImmissionsschutzgesetzBImSchG) last change 10.12.1990

Directives:

- Erste Verordnung zur Durchführung des Bundes-Immissionsschutzgesetzes (Verordnung über Kleinfeuerungsanlagen - 1. BImSchV) of 15.07.1988
- Dritte Verordnung zur Durchführung des Bundes-Im missionsschutzgesetzes (Verordnung über Schwefelgehalt von

leichtem Heizöl und Dieselkraftstoff - 3. BImSchV) last change 14.12.1987

- Vierte Verordnung zur Durchführung des Bundes-Immissionsschutzgesetzes (Verordnung über genehmigungsbedürftige Anlagen - 4. BImSchV) last change 28.08.1991
- Dreizehnte Verordnung zur Durchführung des Bundesimmissionsschutzgesetzes (Verordnung über Großfeuerungsanlagen - 13. BImSchV) of 22.06.1983
- Fünfzehnte Verordnung zur Durchführung des Bundesimmissionsschutzgesetzes (Baumaschinenlärm-Verordnung - 15. BImSchV) last change 23.02.1988
- Zwanzigste Verordnung zur Durchführung des Bundesimmissionsschutzgesetzes (Verordnung zur Begrenzung der Kohlenwasserstoffemissionen beim Umfüllen und Lagern von Ottokraftstoffen)of 7.10.92
- Einundzwanzigste Verordnung zur Durchführung des Bundesimmissionsschutzgesetzes (Verordnung zur Begrenzung von Kohlenwasserstoffemissionen bei der Betankung von Kraftfahrzeugen) of 7.10.92

Gesetz zur Verminderung von Luftverunreinigungen durch Bleiverbindungen in Ottokraftstoffen für Kraftfahrzeugmotore (Benzinbleigesetz -BzBlG) last change 18.12.1987

Gesetz zur Einsparung von Energie in Gebäuden (Energieeinsparungsgesetz - EnEG) last change 20.06.1980

Gesetz über die Einspeisung von Strom aus erneuerbaren Energien in das öffentliche Netz (Stromeinspeisungsgesetz) of 07.12.1990

Annex: Development of emissions

Old Länder

Figure 6

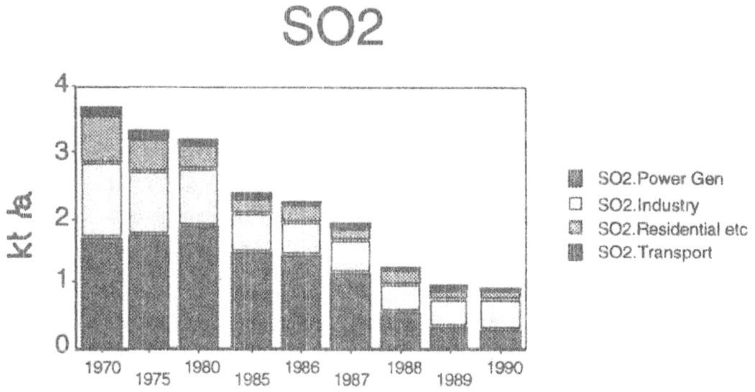

Source: 5. Immissionsschutzbericht der deutschen Bundesregierung, 1992

Figure 7

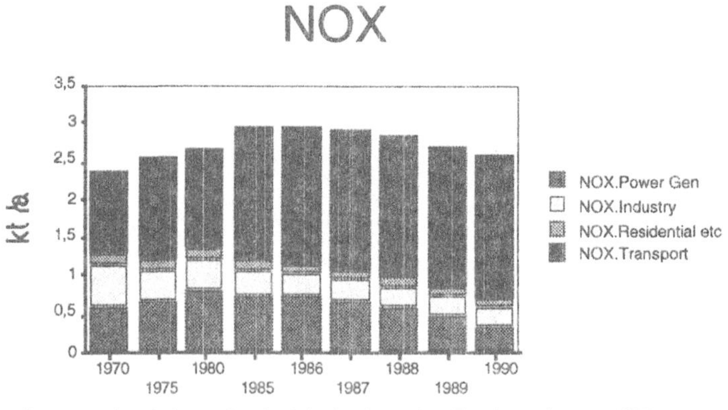

Source: 5. Immissionsschutzbericht der deutschen Bundesregierung, 1992

Figure 8

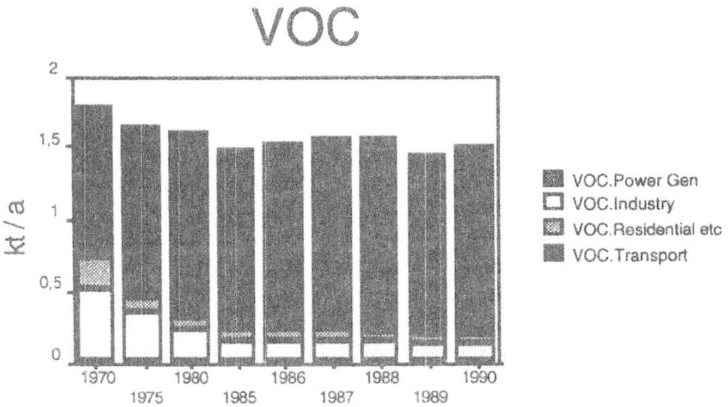

Source: 5. Immissionsschutzbericht der deutschen Bundesregierung, 1992

Figure 9

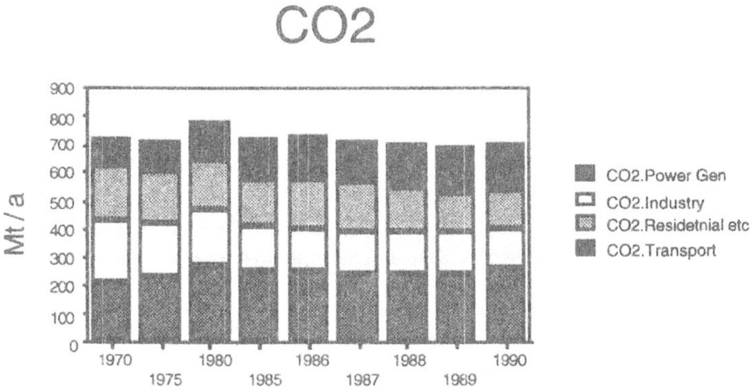

Source: 5. Immissionsschutzbericht der deutschen Bundesregierung, 1992

New Länder

Figure 10

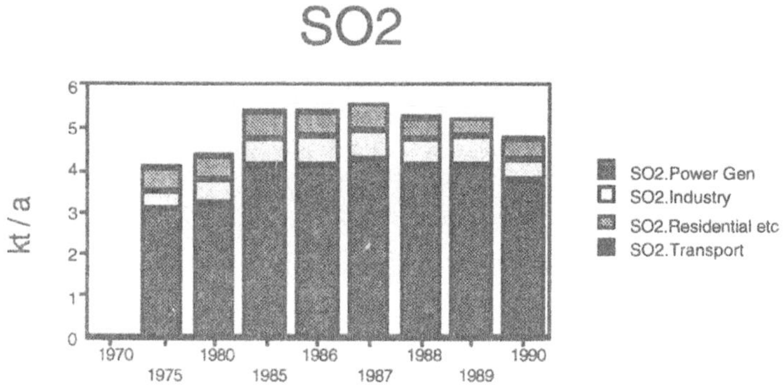

Source: 5. Immissionsschutzbericht der deutschen Bundesregierung, 1992

Figure 11

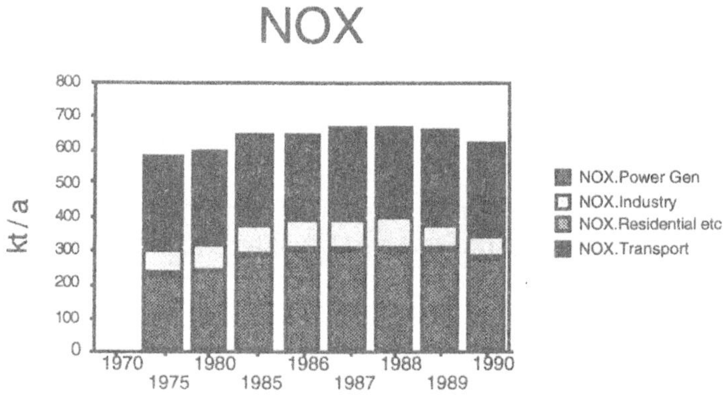

Source: 5. Immissionschutzbericht der deutschen Bundesregierung, 1992

Figure 12

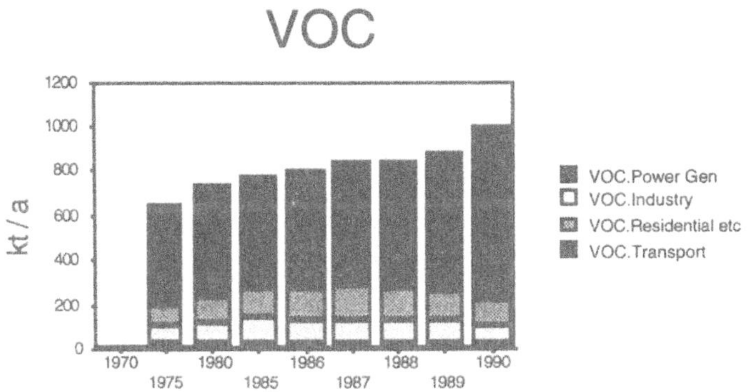

Source: 5. Immissionsschutzbericht der deutschen Bundesregierung, 1992

Figure 13

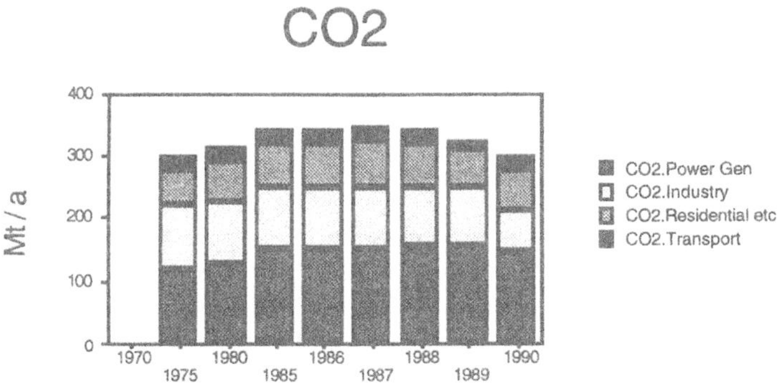

Source: 5. Immissionsschutzbericht der deurtschen Bundesregierung, 1992

Import and export of emissions

Figure 14

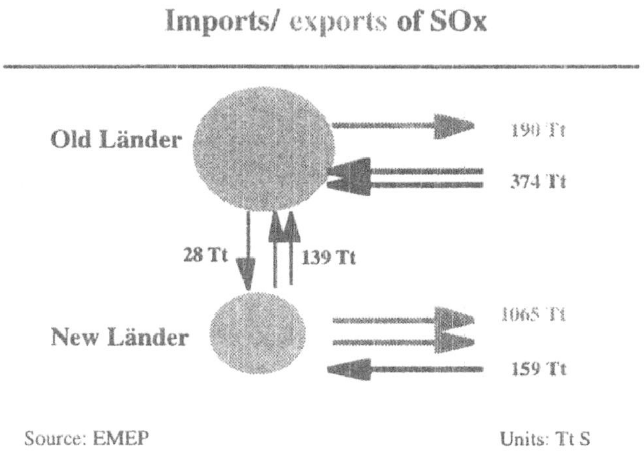

Figure 15

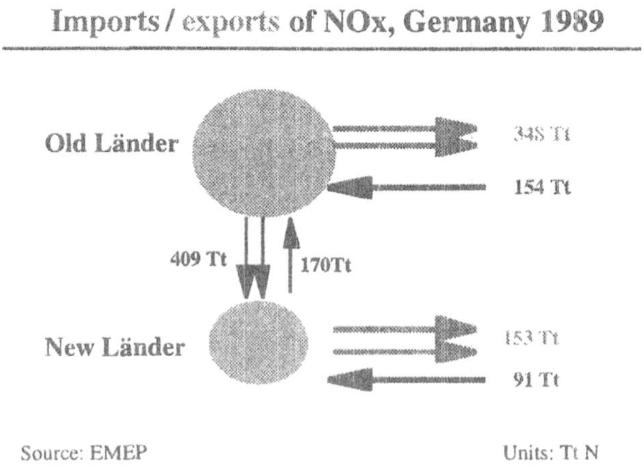

Figure 16

Deposition of NOx and SO2

Germany 1989/1991

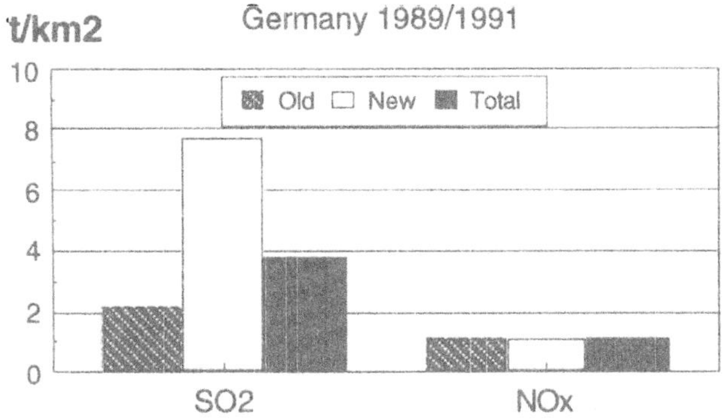

Source: Daten zur Umwelt, 1992 and 5. Immisionschutzbericht 1992

Notes

1 The paper neither deals with the problems of nuclear energy nor does it discuss the problems of harmonization of environmental policy in the EC.

2 Financial support often comes from large electricity producers.

3 See Kornai, J. (1980): Economics of shortage, Amsterdam.

4 Gesetz zum Schutz vor schädlichen Umwelteinwirkungen durch Luftverunreinigungen, Geräusche, Erschütterungen und ähnliche Vorgänge - Bundesimissionsschutzgesetz.

5 Almost all uses of water with the main exception of agricultural purposes according to this law are restricted and require permission. Wasserhaushaltsgesetz, WHG, 12.2.1990

6 §19 g WHG

7 Verordnung über Schwefelgehalt von leichtem Heizöl und Dieselkraftstoff, 3. BImSchV, 15.1.75. This directive contains a gradual increase in the standard from 0.55% in 1975 to 0.2 % valid from 1.3.88.

8 Benzin-Blei-Gesetz - BzBlG, Latest amendment 18.12.1987

9 Verordnung über Chlor- und Bromverbindungen als Kraftstoffzusatz -19.BimSchV, 17.1.1992

10 Verordnung über Kleinfeuerungsanlagen - 1.BImSchV, 15.07.1988

11 For large equipment: Verordnung über Großfeuerungsanlagen - 13.BimSchV, 22.06.83, smaller equipment: Verordnung über.Kleinfeuerungsanlagen - 1. BImSchV, 15.07.88

12 Straßenverkehrszulassungsordnung, StvZO

13 20. and 21. BImSchV 7.10.92

14 §5, section 4 BImSchG.

15 Abfallverbrennungsverordnung, 17. BImSchV, 23.11.1990, particuarly §8.

16 According to a government report in: Umwelt Nr. 7/8, 1993, p. 294

17 Energieeinsparungsgesetz - EnEG, 20.06.1980

18 Gesetz über die Einspeisung von Strom aus erneuerbaren Energien in das öffentliche Netz (Stromeinspeisungsgesetz), 7.12.1990

19 Niedersächsisches Umweltministerium, Leitlinie zur Anwendung der Eingriffsregelung des Niedersächsischen Naturschutzgesetzes bei der Errichtung von Windenergieanlagen, 21.06.1993

20 Details can be seen from: Pfaffenberger, W.: Economic effects of $C0_2$ reduction in Germany, in: Kuckshinrichs, W. u.a.: Workshop on the economics of the greenhouse effect, modeling strategies and impacts, Bad Zwischenahn 1993, p. 15-32

21 Witte, H. u.a.: Umweltschutzmaßnahmen und volkswirtschaftliche Rentabilität, Umweltbundesamt Bericht 4/92, Berlin 1992

22 See Welsch, H.: Kosten der SO_2-Minderung unter alternativen umweltpolitischen Strategien: Ergebnisse eines Simulationsmodells des westdeutschen Kraftwerkssektors, in: Zeitschrift für Energiewirtschaft, 1989, 51 - 59

CHAPTER 13

Problems of multi-regulation: The potential impact of European rules and regulations on the organization of national energy sectors

Leigh Hancher

Introduction

This paper addresses an issue which is of a certain novelty - the impact of Community rules and regulations on the way in which energy sectors are organized and indeed regulated at the national level. I have deliberately chosen to qualify the word 'novelty' for several reasons. Firstly, as is generally well-known, the three Treaties establishing the European Communities (the EEC, ECSC and Euratom Treaties) have, at least in theory, applied to national energy sectors since the 1950s. Nevertheless, many of the powers available to the European Commission, the institution entrusted with the role of guardian of the Community constitution, have never been fully exercised. The reasons for the 'benign neglect' of energy are complex, and can largely be attributed to the sensitive political and economic nature of the sector.

Since 1988, following the adoption of the Commission's Working Paper on the Internal Energy Market, the Community has become more actively involved in matters relating to the organization and regulation of national energy industries, and especially the gas and electricity industries. The 1988 Working Paper comprises in the main of a long litany of the obstacles to the creation of a genuine internal or single energy market and proposes a type of action programme, designed to lead to the gradual removal of these same obstacles.

As part of this action programme, the Commission has now commenced a so-called 'staged' or gradual approach to the creation of the internal energy market. The 'first stage' of the programme has led to the adoption of a series of Directives on transit through high-voltage electricity grids, on transit through high-pressure gas pipelines, and finally on price transparency. The 'second stage' of the process involves further market liberalization, in the form of proposed Directives on common rules for the completion of the internal market in electricity and gas, as well as a concerted effort on the part of the Commission to eliminate unnecessary state subsidies to the energy sector as a whole.

Although this recent initiative demonstrates a rather novel exercise of the Communities' powers as far as the energy sector is concerned, those powers themselves are by no means new. They have been applied in numerous other sectors of the economy, ranging from food to funerals and football, and have been used to regulate activities from marketing to massages.

This is not to deny that the exercise of Community powers over the organization of national energy markets raises a number of specific regulatory problems. These problems are in part to be attributed to the nature of the sector itself and in part to the nature of the Community regulatory process. These problems may be conveniently grouped under the heading 'multi-regulation'. The first part of this paper deals in some detail with the various dimensions of multi-regulation as a general phenomenon. The second part goes on to discuss the current Commission proposals for the electricity market in some detail.

What is multi-regulation?

The nature of community regulation

Before examining the process of regulation in a single sector, some comments on the general Community executive process and the relationship between the adoption and the execution of legislation are in order. In general, the Council, in its capacity of legislative decision-maker, leaves it to the Member State or to the Commission to execute the legislation adopted. The Member States have been described as 'the mainstream executive branch of the Community government';[1] if a legislative act does not provide for any specific mode of execution, it will be for the Member States to take the appropriate measures, whether general or particular, to ensure the effective application and enforcement of the measure in question. There is nothing to prevent the Community legislator from imposing strict conditions on the scope and conditions of exercise of these executive powers, and indeed this is probably the only way in which uniformity in the execution of Community legislation throughout the various Member States can be assured.

This institutional set-up has been described as a type of 'Vollzugfoederalismus' or executive federalism,[2] in which the division of powers between the central government and the component entities is not just one defined in terms of areas in which each government holds substantive competence, but relates also to the division between the central government holding the legislative power and the component entities holding the executive power in a given area. It is within this general framework of executive federalism that the issue of Community regulation and its impact at the national level must be situated.

The recent Community regulatory initiatives in the energy sector raise a number of issues on the relationship between national regulation and Community objectives, and the appropriate division of competence between Community and national institutions. These are discussed further in this paper as issues of multi-regulation.

Problems of multi-regulation in the Community context can be further divided into four broad categories, namely: who should regulate, what should be regulated, what instruments of regulation should be selected and finally, how should those instruments be enforced. These four dimensions will now be explored to identify certain features of multi-regulation as a general phenomenon. Having explored these issues, I shall then consider several aspects of multi-regulation in the concrete context of the Commission's recent regulatory proposals for the energy sector.

Who should regulate?

The issue of the proper delineation between Community and national regulatory competences is by no means a new one. The debate over where

Community competence should begin and end, and the corresponding role of the Member States, has of course been given a new lease of life since the concept of 'subsidiarity' has been incorporated into the text of the Treaty of Union, adopted in Maastricht in December 1991. This in turn raises issues of vertical multi-regulation. The adoption of the subsidiarity principle implies that the regulatory powers of either the Community or the Member States should be confined to a residual category. The question of which level enjoys primary powers and which level may only exercise residual powers is not always an easy one to resolve a priori, as we shall see below. In any event it is quite safe to assume that relationship between Community and national legislation is not a simple hierarchical one.

Community regulation of the energy sector also raises questions of potential conflict between environmental and energy market objectives, and therefore problems of what may be termed horizontal multi-regulation. Following the adoption of the Single European Act a new Title on Environmental Policy was incorporated into the Treaty of Rome (Articles 130R-T). Article 130R(2) states that 'Environmental protection requirements shall be a component of the Community's other policies'. The so-called 'integration principle' has been further strengthened in the new version of Article 130R(2) of the Treaty on European Union (TEU). In its new version, this Article provides that 'Environmental protection requirements must be integrated into the definition and implementation of other Community policies'.

Two separate Directorate-Generals are responsible for energy (DG XVII) on the one hand, and environment (DGXI) on the other. Policies relating to energy and environmental issues must therefore be co-ordinated to some degree between these two Directorates. This aim is of course easier to state than to put into practice.[3] Nevertheless, an independent research report on the integration of environmental policy into other EC policy areas concluded that impressive strides towards integration have taken place recently with the production by the Commission of the Fifth Environmental Action Programme[4] and the first proposal for a carbon/energy tax.[5] This latter has necessitated extensive cooperation between the two DGs, although mainly in the form of ad hoc inter-service consultations,[6] the outcome of which was a draft proposal for the carbon/energy tax in May 1992.[7] Since 1989, joint 'Energy and Environment Council meetings have been convened. However, most of the detailed discussion by Member States of the carbon/energy tax has been undertaken by finance, energy and environment ministers meeting separately within the conventional Council framework.

Thus the issue of who can regulate energy markets is by no means a straightforward one; numerous parties are involved and the relationship between resultant regulatory norms has a vertical and a horizontal dimension

What can be regulated?

The next issue concerns who is entitled to regulate what.

It is of course well known that the Treaty of Rome establishing the EEC, unlike the Coal and Steel and Euratom Treaties, makes no express provision for the adoption of a common energy policy, similar to that provided for agriculture and transport. The new Treaty on European Union - hereafter the Maastricht Treaty - which amends the original EEC Treaty - makes passing mention of 'energy matters' in the new Article 3T of the EC Treaty. It is possible, but not necessarily probable, that the Treaty may be further amended to include a specific article on a common energy policy when it is next considered for revision in 1996. Certainly, the European Parliament has recommended that a separate chapter on energy should be included in the revised Treaty.[8] Thus although the Community has at the present time no express competence with regard to energy policy as such, it nevertheless has an implied competence to adopt measures necessary to remove all obstacles at national level necessary to ensure the creation of the internal market. It may do this either by enforcing the basic Treaty provisions on free movement and competition to require the removal of individual obstacles, or it may adopt general harmonizing measures to ensure all Member States conform to similar, Community-imposed objectives.

The scope of the general Treaty principles

It has never been seriously argued that the basic rules on the customs union (Articles 9-12), free movement of goods and services (Articles 30 - 37; Articles 56 -59) as well as the rules on competition (Articles 85-94) do not apply to the energy sector[9]. Indeed, one of the most famous cases in Community law - Case 6/64 Costa v Enel [10]concerned the nationalization of the Italian electricity industry. The European Court of Justice has since ruled on the application of these rules to the oil and gas sectors on a number of occasions.[11] More recently, the lower court, the Court of First Instance has handed down a number of rulings applying the competition rules to the electricity sector.[12]

It is a well-established principle of Community law, that so long as the Community has not enacted harmonizing legislation, the Member States retain temporary powers in certain situations to impose restrictions on the import and export of goods and services. This temporary competence is not of course to be attributed to the principle of subsidiarity but arises from the original schema of the Treaty of Rome. The Treaty articles on free movement and competition take precedence over national law, and in the event of conflict the doctrine of supremacy dictates that the latter must be set aside[13], unless one of the specific exemptions contained in the Treaty can be successfully invoked. Such restrictions on the operation of the basic principles must satisfy the requirements of the relevant Treaty Articles; they must be objectively justifiable and proportionate to the end sought, and they must not result in an arbitrary restriction on trade. Where restrictions result from so-called 'indistinctly applicable measures', i.e. national rules applying to both imports and domestic products or services alike, the European Court of Justice has devised a number of 'mandatory requirements' which may be relied upon by Member States to justify such measures which nonetheless hinder trade. Once again the national measure in question must be objectively justifiable and must be proportionate.

In competition or anti-trust matters, the dividing line between national and Community competence is rather different. The key question is whether a particular agreement or restrictive practice affects inter-state trade. Community competition law applies only when the interstate criterion is fulfilled. If there is no inter-state trade effect, then national competition law should apply. As the jurisprudence of the Court of Justice and the Commission's decision-making practice makes clear, it is often possible to identify an effect on trade between Member States even where the practices or agreements in question appear to be confined to one Member State alone. As the Commission's decision concerning the arrangements made between Scottish Nuclear and the two Scottish electricity company indicates, Article 85 may apply as long as there is a probability that inter-state trade may be affected.[14]

Given that energy utilities in the majority of EC countries enjoy statutory monopolies which extend over the entire national territory, there is seldom any difficulty involved in invoking Article 86 - which prohibits the abuse of a dominant position in a substantial part of the common market.

Secondary legislation

Between 1990 and 1991 the Council adopted three sets of general harmonizing measures - the Directives on the transit of electricity,[15] and gas[16] and on price transparency for electricity and gas.[17] In September 1990 the Council also adopted Council Directive 90/531[18] on the procurement procedures of entities operating in the water, energy, transport and telecommunications sectors. This Directive is based on Article 57(2), Article 66, Article 100a and Article 113 EEC.[19]

This body of legislation only applies to a few limited aspects of energy market organization, however, and it is indeed for this reason that the Commission has proposed the adoption of further 'second-stage' and eventually 'third stage' measures. Indeed the three specific energy measures mentioned above deal largely with procedural and not substantive issues, while the procurement directive is essentially a framework measure. The adoption of these measures has not displaced the Treaty principles; indeed in the field of procurement, the Court of Justice has frequently drawn upon these very principles to fill in what it perceives to be gaps in the framework, and to broaden the scope of the secondary legislation.[20]

Even although these general and specific powers co-exist, it is necessary to stress that the Community's competence on energy matters is by no means exclusive: as we shall see the Member States continue to retain powers to pursue national energy policy goals.

Environment

Environmental policy is of course expressly included in the Treaty of Rome (Articles 130R-T, as amended in the EC Treaty). Article 130R of the EC Treaty lists a number of objectives to be pursued by Community action relating to the environment but the protection of the environment is not the exclusive competence of the Community. This was made clear in the original version of Article 130R(4) which stipulated that:

> The Community shall take action relating to the environment to the extent to which the objectives referred to ... can be better attained at Community level than at the level of the individual States.

Even if the Community has acted, and adopted a measure under Article 130S, a Member State remains competent to maintain 'more stringent protection measures compatible with this Treaty' (Article 130T).

Maastricht and the subsidiarity principle

The new Article 3(b) EC provides that the Community shall act within the limits of the powers conferred upon it by the amended EEC Treaty[21] and of the objectives assigned to it therein. In the areas which do not fall within its exclusive jurisdiction, the Community shall take action in accordance with the principle of subsidiarity only if and insofar as the objectives of the proposed action cannot be sufficiently achieved by the Member States and can therefore by reason of the scale or effects of proposed action, be better achieved by the Community. Any action by the Community shall not go beyond that what is necessary to achieve the objectives of the Treaty.[22]

The introduction of this ill-defined concept into the Treaty constitutional framework is already the subject of a growing body of literature.[23] The statement published at the conclusion of the Edinburgh summit on the application by the Council of the Subsidiarity Principle is intended to substantiate the rather nebulous concepts set out in Article 3(b).[24] The Council is to incorporate a subsidiarity test in its examination of all the new Commission's proposals. It is probably safe to assume that the principle of subsidiarity should not be understood as a pure demarcation rule assigning exclusive competence to either the Community or reserving it to the Member States.[25] It will be a question of defining which party is best suited to achieving the desired objectives, and by what means.

It is submitted that one way in which the concept of subsidiarity may become important will be in determining the general policy objectives which are to provide a framework for specific regulatory measures. If we take the question of energy policy, we see that the majority of the Member States would appear resolutely opposed to conceding any explicit competence to the Community in energy policy matters, either now or on any subsequent revision to the Treaty on European Union. In the meantime some Commission officials have advocated that if the Commission's energy market liberalization plans were put into a wider energy policy framework, as opposed simply to an internal market one, then its proposals would have to take into account supply-side (including the environment), as well as demand-side measures (including creating an internal energy market). Such an approach would allow for a better integration of environmental policy and energy objectives. In this context, the Energy Commissioner, Mr Abel Matutes, has called for agreement on a common energy policy framework.[26] Pleas in this direction have fallen upon deaf ears at national level in the past.

Given that it is probably unlikely that responsibility for achieving particular objectives can always be allocated between the Community and the national level on a clear-cut, a priori basis, the various norms which can be grouped under the generic heading 'energy regulation' will probably continue to originate at both national and Community levels. In some instances, the national norms may well be either complementary to Community measures, or a further specification of such measures, but in other cases there may well have been no Community action at all so that, subject to the strictures of basic Community law on free movement and competition, it will be up to the Member State to decide on whether to adopt a national energy policy or not, what the components of that policy should be, and what priorities should be set within that policy framework.

To give a concrete example, the Commission has powers to supervise state support to the coal sector under Articles 4, 67 and 95 ECSC. A 'coal subsidies code' enacted in 1986, but which expired in late 1993 provided that aid to the national coal industry can be granted under certain conditions. The Community cannot, however, stipulate whether coal should or should not be produced at national level, it cannot dictate what quantities should be produced, if any, nor does it have any direct influence over whether national coal production remains in private or public ownership.

As we shall see in the following section, ongoing conflicts over who should regulate on such vital matters as security of supply - the Community or the Member States - has played an important role in shaping the scope of the Commission's present proposals on the internal energy market.

What instruments of regulation are to be chosen?

The choice of regulatory instruments is obviously dependent on the aims to be achieved, and this in turn depends on choices of what activities are to be considered legitimate targets of regulation at Community and national level. The complexity and controversial nature of the fundamental policy choices involved here is particularly evident in the energy sector. Essentially, the current debate surrounding the Commission's proposed plans for the internal energy market revolves around the issue of reconciling a market-based approach to energy issues with the more traditional public service-based approaches to energy provision. In those Member States which are traditional supporters of the latter approach, energy market regulation in the form of binding legal norms is almost irrelevant; the public-service monopoly is the market and it usually regulates that market. It is only as a consequence of a deliberate policy of de-regulation that more detailed and direct state regulation becomes necessary.

Public service duties can be guaranteed through regulation, but the importance ascribed to this matter will obviously be reflected in the way in which the regulation is framed. The protection of public service missions may be seen as a predominant or a residual issue at national level, depending on a variety of economic, political as well as geographical factors.

This issue brings me to a further dimension of the concept of multi-regulation at Community level. It is all too frequently assumed that a Community measure is a result of Commission initiatives, prepared and drafted 'in-house' by the latter, to be eventually adopted, usually in amended form, as Council measures. Obviously the Member States participate in this process in Council. But the Member States have an important if less-formalized input at the drafting stage; they are not presented with a Commission fait accompli in Council. Consultations between Member States and the Commission on the present legislative package for the energy sector, for example, have extended over almost two years of negotiations. The European electricity and gas industries have now both formed separate EC-wide associations to press their case at European level. Increased sums of money are devoted by industry to retaining specialists to lobby the European Parliament. Obviously, the more controversial the proposal is, the less are the chances that the Commission's original intentions survive intact. This will have two possible results; fudging and/or deliberate vagueness. Both have obvious consequences for the design of the regulatory instrument in question.

In policy areas where Community regulation is a relatively established feature, such as the environment, the process of concertation can lead to the adoption of a single measure - usually a Directive - which effectively incorporates different types of regulatory instruments, reflecting existing regulatory divergences between Member States. A good example of a 'mixed' instrument is Directive 80/799 on ambient air quality standards which allows quite different tests to be used in measuring pollution levels.

Recent Community environmental legislation offers even wider scope for the adoption of different types of regulatory instruments at national level. The Community directives on car emissions, for example, were adopted in part as a reaction to the various national fiscal incentives for 'clean cars' which were allegedly distorting trade within the EC. The Car

Emissions Directive expressly authorizes national fiscal incentives, but does not seek to harmonize their rates or methods of computation. Instead it provides that these incentives must apply generally and aim at early compliance with the binding Community standards, which had not yet been adopted. Once the latter standards are in place, national incentives must cease to exist.[27]

Where consensus on the best way to pursue environmental protection goals has proved impossible to attain, the Commission is now prepared to incorporate quite separate regulatory policies and instruments within one measure. A good example, is to be found in the recent proposal on waste incineration which incorporates the quite divergent French and German approaches to this issue into a single draft measure.

The recent debate over subsidiarity has also prompted discussion as to whether the Member States should be given more freedom to select the types of instruments with which to pursue Community goals and objectives. In principle, the most widely used instrument of Community regulation, the directive, already confers considerable discretion on Member States in their choice of method of implementing the Community directive in question.

How is regulation to be enforced?

It is probably in the sphere of regulatory enforcement that the inter-play of Community and national regulation has long been the most evident. At the Community level, and outside the field of competition law, the possibilities for enforcement are limited; this task is left to the competent authorities of the Member States themselves. In addition, private individuals and companies may enforce their Community legal rights in their national courts, and the latter are obliged to give effect to those rights. Sanctions, such as damages for financial loss, for breach of Community norms are imposed at national, not Community level. Criminal sanctions and penalties must also be imposed through national processes.

Even in terms of monitoring national enforcement practices, the Commission has limited resources. In recent years efforts have been made to reinforce this monitoring function. In some cases this has led to the establishment of specialized Community bodies, but such Community agencies, such as the European Environmental Agency are the exception, not the rule, and even then their monitoring powers are limited to information gathering.[28]

In conclusion, and to emphasize a point already made, we cannot treat the issue of multi-regulation in the Community context as a simple hierarchical process, where regulatory norms stipulated at Community level will displace national norms, while leaving Member States some discretion in the choice of instruments for their execution or implementation as well as responsibility for their enforcement. The process of multi-regulation must be seen as a complex multi-dimensional process, with different levels of input and output.

Multi-regulation in the energy market

I shall now turn to examine the issue of multi-regulation and the Commission's recent proposals for energy market liberalization. In the following sections I shall first look at the different conceptions of the application of the subsidiarity principle to the energy sector. This, I shall argue, has generated debate not only about whether the Community should intervene in certain aspects of energy market legislation, but how it should intervene. Finally, I shall turn to the types of regulatory instruments which are proposed and discuss some of the problems which they raise.

Internal energy market initiatives and subsidiarity

The two draft Directives on gas and electricity, first proposed in February 1992 above, are seen by the Commission as a concrete expression of the concept of subsidiarity in the regulation of the European energy sector. This is primarily reflected in the nature of the proposed instruments; the proposed rules are drafted as framework rules, to be further specified at national level. If and when these two proposed Directives are eventually adopted in Council, problems of what I have termed 'vertical' multi-regulation are certainly likely to emerge in their implementation. At the same time, the coverage of the proposed rules is limited to certain matters; thus problems involving horizontal multi- regulation cannot be ruled out in the future.

These proposals, designed to achieve the second stage of the construction of the internal energy market, are more ambitious than the earlier transit directives. They are based on the following objectives:

- the introduction of competition in electricity generation and in gas and electricity transmission;
- the separation or 'unbundling' of various functions performed by vertically integrated companies;

- the introduction of a limited degree of third party access.[29]

As regards the production of hydrocarbons, a separate proposal for a Directive on the conditions for granting and using authorizations for the prospecting, exploration and extraction of hydrocarbons was presented to the Council on 11 May 1992.[30] A common position was adopted on 16 December 1993. The Directive's objectives are to ensure undertakings equal access to oil and gas resources and free movement of these products between Member States.

There are clearly a number of approaches to interpreting the application of the subsidiarity principle to the Community energy market as the debate and discussions which have surrounded these controversial proposals clearly illustrate.

The Commission's approach

In the explanatory memorandum accompanying its original proposals published in February 1992, but drawn up prior to the finalization of the relevant new EC Treaty Article 3b at Maastricht in December 1992, the Commission took a particular approach to the subsidiarity issue. The principle was described as one of four general principles informing its entire approach to the gradual liberalization of the energy market. Subsidiarity required that the Community must not impose rigid mechanisms, but rather should define a framework enabling Member States to opt for the system best suited to their natural resources, the state of their industry and their energy policies.

Thus the Member States should remain free to determine, in an objective and non-discriminatory manner, the criteria for the granting of licenses for the construction of power stations, cables and pipelines. In the case of electricity, they can also specify the criteria for dispatching power stations. They may also continue to regulate all aspects of electricity and gas tariffs for all final customers not eligible for third-party access or not opting for it. This includes the possibility of tariff equalization at national level, although cross-subsidization from 'reserved' markets to markets under competition is not permissible. The Member States are also to remain free to determine the extent and nature of the rights and obligations of firms supplying gas and electricity, in particular their public service obligations. They may also continue to grant distribution companies exclusive supply rights, and to impose the obligation to connect and supply all the final consumers not eligible for third party access. Finally, each Member State is left to choose the manner in which the directives are to implemented, that is either by a specialized regulatory authority, or though the general mechanisms of national competition law.

The European Parliament

The Parliament formally considered the Commission's proposed package, following a public hearing in October 1993. In the view of Mr. Claude Desema, the European Parliament's rapporteur on the 1992 package of directives, subsidiarity must constitute the guiding principle with regard to both harmonization and liberalization: '[I]n other words, Member States will retain exclusive responsibility for rules and provisions that are not laid down in these or subsequent directives'.[31] The European Parliament's Rapporteur has taken issue with the very approach to market liberalization set out in the Commission's original proposals. In his view the Commission should place an emphasis on harmonization over and above liberalization, the latter being conditional on the former. In particular his report advocates that further harmonizing measures are necessary on essential matters such as environmental protection requirements, the rules governing taxation on production, transmission and distribution of electricity and natural gas, rules on transparency of costs and the definition of public service obligations and implementing rules. Any further development of the energy system towards further integration and thus towards further liberalization will necessarily be linked to the implementation of a common energy policy, based on provisions to be inserted into the Treaty on European Union when it is revised in 1996. In the meantime, and until progress on harmonizing measures is secured, Member States should be free to organize their energy markets as they see fit, subject to the minimum rules laid down in the amended version of the present proposals.

In its Opinion on the two proposals, the Economic and Social Committee (ECS), also considered the application of the subsidiarity principle. It took a slightly different view, and it has linked the concept of security of supply with that of subsidiarity. It observed:

> 'Security of supply is not seriously considered in the proposed Directives. It is not included in the list of definitions in Article 2 .... This is a grave defect in these Directives; it must be clearly stated that such security of supply is primarily the responsibility of the distributors of natural gas and electricity and this responsibility must be underpinned by proper institutional arrangements in each Member State, in accordance with the principle of subsidiarity.

According to the principle of subsidiarity this security of supply shall be guaranteed by the States themselves, combined with the necessary Community-level co-ordination (paras. 2.2.2 - 2.2.3.).

In the remainder of its report, the ECS goes on to criticize the two draft directives for failing to provide Member States with sufficient guidance on what security of supply means and how it should be guaranteed in a Community context. The Opinion concludes by supporting the Commission proposals in limited respects only, and in particular rejects the plans for the introduction of third party access (TPA). It too calls for greater emphasis to be placed on an alternative approach, based on harmonization of environmental rules and the development of a balanced and gradual approach to common commercial criteria and common concepts of security and quality of supply.[32]

The energy utilities

The majority of the national utilities in the gas and electricity sectors have opposed the Commission's plans, although they have been prepared to accept that some form of change is inevitable. The C.E.E.P., the public sector enterprises' representative body in Brussels, has responded to the Commission's plans to introduce more liberalization into a number of utility sectors by floating the idea for a European Charter of Public Service. This Charter would set out and define the role of public firms in the provision of utility services. In particular the C.E.E.P. also argues that the principle of subsidiarity should be applied in such a way that Member States should be allowed to choose how to manage their public services, except when these services are located at EC level.[33]

The operation of the principle of subsidiarity in relation to energy policy matters is therefore very differently conceived by the Commission, the Member States, the European Parliament and the energy utilities themselves. Whereas the Commission assumes it has competence to enact energy-market related legislation, it must observe certain criteria in the implementation of that legislation. This in turn has consequences for **how** energy markets are to be regulated; what aspects may be regulated and by what means? Many Member States and their utilities dispute the Commission's right to intervene in energy policy matters at all, while the Parliament seems to assume that the principle requires that the Commission should only assume competence for matters relating to energy market liberalization if and when the relevant national rules have been harmonized. Until this point in time, the Member States should retain exclusive competence to organize their national markets as they choose.

These different interpretations of the concept of subsidiarity are certainly of considerable relevance to the essentially political aspects of the debate over the desirability of Community intervention in energy market matters. They would appear to be of somewhat less relevance to the legal limitations, if any, which the principle of subsidiarity imposes on Community legislative competence in the energy sector.[34]

How is the energy market to be regulated

Turning now to the second main issue with which this part of the paper sets out to deal - the question of how the internal energy market is to be regulated - it should be pointed out that here we are confronted not only with regulation by secondary legislation but also via the application of the relevant Treaty articles on free movement and competition which the Commission has autonomous powers to enforce; secondary legislation is not required. Subsidiarity, in the sense just discussed, is primarily concerned with allocating legislative competence. It should also be noted that national courts also have a role to play in the enforcement of Community law. Those Treaty articles, as well as relevant provisions of secondary legislation which have been or are capable of being designated as directly effective may be relied upon by private individuals in their national courts, and the latter are under a duty to apply the relevant principles of Community law, if necessary setting aside conflicting provisions of national law.

Conduct and structure regulation

The Treaty rules on competition are designed primarily to regulate the conduct of undertakings and to some extent, Member States, while secondary legislation is needed to deal with structural issues. As Vickers has suggested, 'structure regulation' determines which firms or types of firms can engage in which activities, while 'conduct regulation' concerns the permitted behaviour of firms in their chosen activities. Thus the secondary legislation proposed for the energy sector is largely structural; it seeks to determine the types of firms which may engage in energy production, transportation and distribution, and lays down rules, such as unbundling, on how the firms should be organized.[35] In other words, the draft legislation is primarily a deregulation measure.

Nevertheless, the Treaty rules on competition are essentially limited to controlling certain broad categories of conduct which is deemed **ex post facto** to be anti-competitive. Conduct regulation - in the form of secondary legislation - is still necessary - to specify in detail and **a priori**, the standards which must be observed in particular sectors.

The draft directives on the creation of the internal market also contain conduct rules. To give a concrete example, the proposed electricity directive requires the designation of a transmission system operator (TSO), whose activities must be restricted to supervising transmission. Energy production and energy trading must be structurally separated from energy transportation. Nevertheless, this is not sufficient to guarantee that the TSO, which will after all remain a national or at least a regional monopoly, will deal with all its potential customers fairly. The draft directive therefore includes rules relating to pricing, the drafting of technical standards on interconnection etc. Similarly, although production and distribution must be 'unbundled', the supply of electricity to certain categories of customers may remain reserved to national or regional firms. Hence conduct rules on cross-subsidization are required.

The present package of measures therefore contains elements of de-regulation as well as re-regulation; they include elements of structural as well as conduct regulation. Interestingly, the draft proposals on the common rules for completing the internal market for electricity and gas were preceded by two separate sets of proposals, prepared by the Commission services in 1991. It was originally envisaged that the Commission would adopt two Directives based on Article 90(3) EEC, requiring Member States to withdraw exclusive rights to import, export, produce, supply and market electricity and gas respectively.

Article 90(3) is in fact one of the few Treaty Articles which grants the Commission power to adopt legally binding rules in furtherance of its supervisory powers over statutory monopolies and other types of regulated firms, without obtaining the support of the other Community institutions. The two proposed draft Directives would be complemented by two Council Directives, based on Article 100A, introducing certain common rules for the regulation of these two markets. In this approach, the Article 90(3) Directives would have implemented the structural aspects of the Commission's proposed strategy, while the conduct rules would have been incorporated into Council Directives.

This approach essentially mirrored that already taken in the field of telecommunications.[36] Two Commission Directives, based on Article 90(3), had been adopted requiring the withdrawal of exclusive rights to supply respectively, telecommunications goods[37] and services and thus an extensive restructuring of national telecommunications authorities, while separate Council Directives deal with conduct-related problems, including conditions of access for competitors to networks.[38]

The Commission's power to adopt the two Article 90(3) Telecommunications Directives was subsequently challenged by several Member States, but confirmed in both cases by the Court of Justice.[39] Although the Court's rulings would appear to have given the 'green light' to the Commission to adopt future measures based on Article 90(3), the Commission has indicated that for political reasons it will base energy market liberalization measures on Article 100A.[40] Nevertheless, the closing recital of each of the 1992 Draft proposals notes that 'the Commission reserves the right to make use of all its powers conferred on it by the Treaty as and when appropriate'. Thus future recourse to Article 90(3) is not entirely precluded.

In the meantime, the Commission has begun infringement proceedings against six (originally nine) Member States, on the basis of Article 169 EEC, requiring the removal of import and export monopolies for electricity and gas. The Commission argues that these statutory monopolies infringe the Treaty provisions requiring the abolition of all measures equivalent to quantitative restrictions on imports and exports (Articles 30 and 34) and the related provision requiring Member States to adjust their state trading monopolies (Article 37). The Commission has a discretion as to whether or not to proceed to the second or judicial phase of the Article 169 procedure, which involves seeking a declaration from the Court of Justice that the Member State in question has failed to comply with its obligations under the Treaty. The Treaty on European Union makes provision for the Commission to request the Court to impose fines in the event of non-compliance with a ruling.

The Commissioner for Competition initially indicated that he is prepared to reconsider making an application to the Court of Justice if the Member States concerned are willing to co-operate with the Commission on the adoption of the common rules.[41] This recent incident offers a useful illustration of the interaction between the different sets of regulatory instruments available at Community level - the general and the particular. It is obviously more useful to the Commission to secure agreement on the adoption of relatively detailed conduct rules than to risk alienating Member States by legal action which will result only in an adjustment (but not abolition) of the structure of their national monopolies. Had the Commission pressed ahead with its original proposals for the adoption of the two Article 90(3) directives, which were also essentially aimed at structural alteration - the removal of exclusive rights and privileges - it would have risked losing all support for the complementary proposals which would have imposed the necessary conduct rules. As matters have turned out, the Commission has now decided to commence infringement proceedings against six Member States in respect of their statutory import or export monopolies.

The instruments proposed

As already indicated, the 1992 version of the draft common rules are intended to secure several objectives: competition, unbundling and transparency and a limited degree of market access. The instruments to achieve each of these various objectives will now be examined.

The original proposals for common rules on competition in the Community's gas and electricity markets are divided into 6 chapters. I shall briefly outline the content of each and then examine their regulatory implications.

Market access and competition in production

Chapter I (Articles 3 et seq) deals with access to the market. Member States shall ensure that electricity and gas undertakings are operated on commercial principles and shall not discriminate between these undertakings as regards either rights or obligations.

They shall allow electricity undertakings established in the Community to build, operate, purchase or sell generating installations which are located on their territory and which are intended for the generation of electricity destined for own use or for sale subject only to objective and non-discriminatory criteria and procedures established in accordance with the Directive. These criteria shall relate exclusively to:

- security and safety of the installation
- environmental protection
- land use and siting
- the technical and financial capacity of the applicant.

These may be referred to as the basic licensing criteria.

The Member State may for reasons of environmental policy or security of supply supplement these criteria by criteria restricting the nature of the primary energy source that may be used for the generation of electricity (Articles 4 and 5).[42]

The gas proposal provides that the Member States shall grant licenses to build or operate LNG facilities and storage facilities, subject to similar criteria (Article 4).

Transmission licenses, and licenses to build or operate distribution lines/pipelines must also be granted on an objective and non-discriminatory basis, and should be granted only subject to the basic criteria, as above. A refusal to grant a licence can only be justified where requirements can be satisfied by the existing transmission and distribution capacities available in the interconnected system at a reasonable and equitable price.

Member States shall ensure that electricity and gas producers and suppliers established in their territory are able to supply their own premises, subsidiaries and affiliates and customers in the same or other Member States, either through a direct line (Article 6) or through the public grid network (Article 7). Customers have the right to buy from a producer of their choice, irrespective of its location and to have the electricity/gas delivered through a direct line or through the public network.

In the case of electricity, a Member State may, if it chooses, limit access to the public system:

- to companies for the supply of sites the overall consumption of which exceeds 100GWh;
- to distribution companies, individually or in association, whose individual or aggregated sales represent at least 3% of overall consumption in the Member State concerned (Article 7(2)).

In the case of gas, access may be limited to:

- individual companies whose overall consumption exceeds 25 million cubic metres per year; and
- distribution companies, individually or in association, whose individual or aggregated sales represent at least 1 per cent of overall consumption (Article 6).

Transmission system operation

In the case of electricity, the Member States will be required to designate or require undertakings which own networks to designate a TSO. The Commission is of the opinion that the structure of the gas market is such that there is no need to designate separate, centrally designated gas transmission operators. As gas, unlike electricity can be stored, and pipeline pressure varied, any network operator can be responsible for the operation of its own part of the pipeline.

Chapter III (Articles 8 et seq.) deals with transmission and the TSO. This chapter requires administrative unbundling for electricity (Article 8(2)). The transmission operator will also be forbidden from buying or selling electricity, except in a number of limited circumstances. An example given by Article 13(4) is the requirement that the transmission system operator shall give priority to generating installations using renewables or alternative energy sources with a capacity of less than 25MW. Thus it would seem that transmission system operators could be compelled to purchase this type of energy.

The operator will also be under a series of obligations, to facilitate energy transfers, to conclude agreements with other transmission or distribution systems to enable use of the interconnected system and to produce detailed information on the operation of the system (Articles 9 and 10). In accordance with Article 12, the TSO must prepare an annual estimate on generation capacity and demand for electricity, covering a period of at least 10 years. Within one year of the entry into force of the Directive, the operator will be required to develop and publish Technical Rules. These rules must be approved by the Member States and notified to the Commission in accordance with the general Community rules on notification of new standards and specifications (Article 12).

Article 13 governs dispatch. The dispatch of generating installations and the use of interconnectors shall be determined by the TSO according to the actual needs of the system on the basis of criteria which are approved by the Member State concerned. These must be objective, transparent, and applied in a non-discriminatory manner. Further, they must not disturb the functioning of the internal market for electricity. A Member State may, however, require that priority be given to the dispatch of generating installations using indigenous fuel to an extent not exceeding 20% of the overall primary energy necessary to produce the electricity consumed in the Member State concerned. This figure shall be progressively reduced to 15% by December 2000.

System users and prospective users may apply to the operator to enter into an agreement for connection to and/or use of the public grid system and the latter will be under an obligation to propose such an agreement. Refusal is only allowed on the grounds related to the fulfillment of any statutory obligation or contractual commitments. A response to an application should be given within three months following its receipt (Article 14(2)).

For gas transmission, the obligations on the transmission companies which individually assure the operation of the integrated grid system are essentially similar (Article 7 et seq).

Distribution

In accordance with Chapter IV (Article 16, electricity; Article 14, gas) Member States may oblige distribution companies to serve particular classes of customer located in a given area. They must also designate a distribution system operator which will have the duty to operate and ensure the maintenance, system reliability, efficiency and further development in its area. The distribution operator will be obliged to conform to similar types of rules on access, communicating technical and financial information, etc., as the transmission operator (see Articles 17 to 22 electricity; Articles 15 to 20 gas). The distribution system operator will also have to produce an annual report on the quality of supply and service, in the form of a standardized report, as prepared by the Commission. This report must be made available to neighbouring authorities, to the Member State, and to the EC Statistical Office.

Member States may also regulate the tariffs for fuels supplied to franchise customers, although this power is stated to be without prejudice to the principle that electricity and gas undertakings should be able to operate on a commercial basis (Article 3). Nevertheless, Member States should be able to require standardization of prices at national level.

Those consumers who are not eligible for, or choose not to exercise their TPA rights, have a right to be supplied by their local supplier. This obligation should probably also apply to supplies made to consumers who use a direct line supplier, or access to the grid, for part of its needs only.[43] If a consumer does choose to exercise its TPA rights or chooses another direct line supplier, provision is made for the minimum notice periods of termination (not to exceed three months) and for the consumer to able to request reconnection (not to exceed six months).

Unbundling and transparency of accounts

Chapter V deals with this issue. The Member States will be obliged to ensure that vertically integrated firms organize their electricity generation or gas production, transmission and distribution activities into 'as many separate divisions as there are activities'. In addition the separate divisions must keep separate accounts in accordance with a minimum number of rules laid down in the directive.

Chapter VI contains the final provisions. Article 25 (Article 23 for gas) allows Member States to take emergency measures in the event of a sudden crisis 'in the energy market' - it is not clear whether this is a reference to the national or the Community energy market - or where the physical safety or security of persons, apparatus or installations or system integrity is threatened. The Member State must communicate these measures to the Commission and to the other Member States. The Commission may decide that the Member State concerned shall amend or abolish such measures in so far as they distort competition and adversely affect trade to an extent common to the common interest.

Finally, Member States will be obliged to set up dispute resolution procedures by which the parties can settle a dispute on matters covered by the Directive. They shall also be obliged to set up consultative committees enabling system users including domestic customers to be consulted, at least annually, on matters relating to transmission and distribution in the national territory.

Regulatory implications

The following commentary deals primarily with the proposal on electricity. I will begin with a few remarks on competition at the production stage, then go on to look at transmission and distribution.

Production

First of all it is noteworthy that the Draft Directive makes no reference to Article 222 of the Treaty. This provision, which has substantially similar counterparts in the ECSC Treaty and the Euratom Treaty, provides that:

> This Treaty shall in no way prejudice the rules in Member States governing the system of property ownership.

In other words, public and privately-owned enterprises may co-exist. In accordance with the European Court's established jurisprudence on the relationship between this Article and other Treaty provisions, the Commission must respect the principle of equal treatment and non-discrimination between public and private firms. Furthermore, the mere fact of ownership of certain assets or property rights, whether public or private, does not shield an undertaking from the application of the competition rules when it comes to the use to which these assets or property rights are put.

It is interesting to speculate how this provision may affect the application of the proposed obligation on Member States to allow all undertakings established in the Community to build and operate power stations (Articles 4 and 5). Presumably those countries in which state-owned enterprises enjoy a virtual monopoly over production - for example, France and Greece and, until now, Italy - will now be required to open generation to full competition. Previously only marginal competition from auto-generators has been tolerated in limited circumstances.

Even if the eventual adoption of the Directive might oblige Member States to withdraw monopoly rights over production from national enterprises - whether public or private - it will not necessarily mean that these large monopolies are going to disappear. The Directive is merely designed to facilitate entry. It lays down certain exclusive criteria which Member States may incorporate into their licensing regimes. It is interesting to note that these criteria would not appear to include restrictions imposed in the interests of energy planning. In a number of Member States it is a condition of the licensing regime that a potential generator must have a certain capacity at its disposal, or is planning to generate a certain capacity before it may obtain a licence for public supply. Such restrictions could be justified under criteria relating to the technical and financial capacity of the applicant.

Similarly, criteria in licensing regimes or accompanying arrangements prohibiting per se producers from expanding their activities downstream or distributors from expanding upstream, will be prohibited. This may have the paradoxical effect that those countries which have recently tried to introduce vertical disaggregation into their electricity by means of such prohibitions, will soon have to withdraw them. This may in turn lead, in the longer term, subject to the operation of the unbundling requirements, to the development of regional or even national monopoly power.

In the shorter term, we should not lose sight of the 'escape-clause' for Member States to be found in Article 4(2), which allows Member States, for reasons of environmental policy or security of supply, to supplement these criteria by restricting the nature of the primary energy source that may be used for electricity generation. An argument for nuclear power could be made out on environmental as well as security grounds. Indeed the Commission has itself expressed a commitment to nuclear for these very reasons. Running a nuclear power plant is a costly business; facilities for disposing of radwaste or for decommissioning power plant do not exactly come at bargain-basement prices. It is a condition of most national licensing regimes that adequate provision for such activities is made in advance by the applicant. [44]

Given that the terms 'environmental policy' and most notably, 'security of supply' are not defined, this would appear to leave Member States ample scope to determine not only policy objectives but also the instruments by which this policy is to be implemented.

It is therefore contended that Member States may conceivably find a number of ways to design and operate their licensing regimes in such a way as to guarantee the established position of the state sector firms. Thus the Community rules will not displace national licensing regimes. Instead the situation will be characterized by horizontal multi-regulation with some vertical integration of national licensing provisions into the Community imposed criteria.

In the final event, given that investment in generating plant is a costly business which requires long term planning, competitive entry into the electricity generation market will only take place if this is likely to be economically attractive in the long as well as the short term. Investment decisions by new entrants will be dependent not only on the potential to compete in generation costs with established utilities, but also on their ability to transport their production to their customers on terms and conditions which are reasonable, stable and predictable. Are the proposed rules on transmission likely to meet these tests?

Transmission

An obvious precondition for fair conditions of transmission is that existing, integrated utilities cannot manipulate decisions on prices and capacity in their own favour. This is why unbundling has been proposed. Unfortunately the precise scope of Article 8 is not wholly clear. Although it might be contrary to Article 222 EEC, which guarantees national property regimes, for the Commission to require outright changes in ownership, it is not certain whether the type of 'operational' unbundling envisaged would ensure the complete commercial independence of the TSO from the production or distribution 'arms' of a vertically-integrated company .

Obviously, the development of TPA will depend on the types of rules which the TSO is ultimately subject to, and on the commitment to their enforcement at national level. It is perhaps to be regretted in this context that there is no clear obligation on Member States to enact rules which will ensure that either transmission networks or distribution networks are planned in order to meet the requirements of eligible TPA users. The absence of such an obligation may be especially problematic in areas where transmission lines or pipelines are already used to full or near full capacity. The fortunate owner of these facilities will be in a position to exact a lucrative price for their use. It is exactly to counter this type of situation that the FERC in the USA has required the construction of additional capacity.[45]

The proposed transmission rules (for both gas and electricity) are by no means exhaustive. In the first place, while there is an obligation to respond to a request for transmission within three months, no time limit is fixed for the period in which actual connection should take place. Finally, there is no guidance on the duration of transmission contracts. An independent generator needs to be assured of access over a long period if investments are to be worthwhile. There would appear to be nothing in the proposal to prevent a transmission operator from concluding only short-term contracts.

In the second place, the provisions on pricing are particularly vague and unsatisfactory. Transmission prices must be:

> 'reasonably related to the long-term costs incurred in the provision of the relevant service, together with a reasonable rate of return on the capital employed in the provision of that service' (Article 21(6) electricity)'.

The pricing of access to fixed network facilities raises innumerable questions. It is quite conceivable that different Member States could adopt quite divergent implementing rules and yet remain within the framework of Article 21(6). But it should of course not be forgotten that there is no obligation as such on the Member State to regulate this matter in the first place - they may choose to leave it to the parties concerned to negotiate transmission prices for each individual transaction.

In the third place, provisions such as Article 14 of the electricity proposal which requires transmission system operators to refuse to make a proposal for a connection agreement 'if such use would prejudice the transmission of electricity in fulfillment of any statutory obligation or of contractual commitments', appear to leave ample scope for Member States as well as the existing industry players on the market to tie up spare capacity for any number of years.

In the fourth place, there is no obligation to provide stand-by or reserve supplies to independent generators, nor is there any obligation to purchase surplus power. These types of guarantees are usually considered necessary to encourage independent generation.

The 1992 Draft Directives only require Member States to provide for a form of dispute settlement in cases of disagreement. Yet the Draft Directives oblige both the transmission and the distribution operators to produce a series of reports, draft technical rules, publish statements of opportunity, pricing schedules etc. To what extent, if any will these activities be regulated and by whom?

Indeed the Commission remains generally agnostic as to the type of measures or regulatory regimes which the Member States themselves will be obliged to adopt in order to implement the rules. Article 25 of the 1992 Draft merely requires the Member states to establish mechanisms for dispute resolution. This seems to allow for a number of options, ranging from no more than the type of procedures envisaged in the present Directive on Transit of Electricity through High Voltage grids, right through to a fully-fledged regulatory authority on the lines of the British OFFER.

Finally, and once more in deference to the principle of subsidiarity, the proposals leave it to each Member State to choose how it will enforce the common rules, either by setting up a specialized regulatory agency, or by applying the general national rules on competition law.

The Commission has repeatedly stated its intention to avoid excessive regulation and not to go beyond what is strictly necessary in order to achieve liberalization, and as in the case of transit, to leave ample scope for commercial negotiation between the undertakings concerned. In the light of the controversy that has surrounded the proposals in their current form, it is likely that several of the draft articles considered here, and especially those designed to deal with conduct regulation will be amended, leaving more rather than less scope for national regulation in the future. Even if the proposed secondary legislation is substantially reduced in scope, however, we should not forget that the primary Treaty rules on competition and free movement may still be relied upon to control certain forms of individual conduct in the energy sector. Once again it would be difficult to sustain the conclusion that national regulation will be subsumed into a Community framework; the Community rules are both incomplete and framed at a very general level.

The Commission's amended proposals

The original draft directives, as proposed to the Council in February 1992 have met with enormous criticism, not only from the majority of the Member States and their utilities but also from the other EC institutions. The reactions of the European Parliament and the Economic and Social

Committee have already been discussed. As a result of its inability to persuade the Council to consider its drafts in their original form, the Commission produced an amended version in February 1994, taking on board some, but by no means all of the European Parliament's proposed amendments. In the latest 1994 version, the Commission has sought to defer even more to the Member States, leaving them with yet more scope to define essential aspects of security of supply as well as the nature and scope of the public service obligation which they place upon electricity and gas suppliers. In general the rules have been recast so that there is less emphasis on Community-imposed regulation and more emphasis on Member State choice.

The new proposal expressly acknowledges that 'security of supply and consumer protection entail a number of obligations of general interest which free competiton, left to itself, cannot be relied upon to fulfil'.[46] Implicitly this means that Member States, in deference to the concept of subsidiarity, will retain considerable freedom to decide how they organize their energy sectors as well as how they regulate them.

In particular, the original proposals on unbundling have been redrafted, so that Member States will only be required to ensure administrative unbundling and the production of separate accounts. Mandatory third party access has now been abandonned in favour of non-compulsory or 'negotiated third party access', while the Member States remain free to determine not only public service obligations which they impose at all levels of industry, but also the extent of the monopoly which distribution companies can continue to enjoy over supply to domestic or tariff customers.

Conclusion

In this paper I have attempted to demonstrate that the concept of multi-regulation in the Community context is a multi-faceted or multi-dimensional one. The interaction of Community and national regulation cannot be conceived in purely vertical terms. I have argued that it is necessary to approach the topic of multi-regulation by considering not only who can regulate, but also by focussing on what is to be regulated, and by what instruments and by what means of enforcement. The Commission proposals on the internal energy market are not only limited, vague framework measures, but they are also restricted in their scope and coverage. They deal with some but not all elements of energy market structures and propose general, framework rules to regulate certain forms of conduct in these markets. Member States remain fully competent to set the terms of their own energy policies, to determine the extent and nature of the 'public service obligations' imposed on their utilities, and retain a large measure of discretion to pursue their chosen regulatory strategies to implement these policies. In their latest version, the Commission proposals will have little impact on the structure of national energy markets. The impact of the proposed directives on what can continue to be regulated at national level is likely to be relatively marginal.

As regards the types of instruments which are to be adopted, the common rules deliberately leave substantial discretion to the Member States, not only to the content of the rules to be adopted but also as to their form and most importantly, to choose the means by which the rules will be enforced. This is especially true of the rules which would regulate conduct. One might legitimately ask whether the proposed rules would after all lead to far-reaching change in the organization of national energy markets and their regulation. I would venture to suggest that the only reply which could be given with any certainty is that it will depend on the Member State in question.

Notes

1 K. Lenaerts, *Regulating the regulatory process: delegation of powers in the European Community* (1993) European Law Review, pp 23-49.

2 J. A. Frowein, *Integration and the Federal Experience in Germany and Switzerland* in M. Cappelletti, M. Seccombe and J. Weiler (eds), Integration Through Law, Berlin, De Gruyter, 1986, p586,

3 As long ago as 1979, the Spierenburg Report identified the Commission's excessive compartmentalization 'as one of the greatest drawbacks of the present administrative structure". EC Commission, Proposals for the Reform of the Commission and its Services, Report made by an Independent Review Body under the Chairmanship of Mr. D. Spierenburg, Sept. 1979.

4 EC Commission, *Towards Sustainability*, 1992.

5 Commission document Com(91) 1744, entitled *A Community strategy to limit carbon dioxide emissions and to improve energy efficiency* was issued jointly by the Commissioners responsible for Environment and Energy.

6 A description of the inter-service consultations is to be found in House of Lords' Select Committee on the European Communities, Eighth Report, Session 1991-92, Carbon/Energy Tax, at p. 37.

7 Com (92) 226.

8 See the Introduction to the Report prepared by M. C. Desema, rapporteur to the Committee on Energy, Research and Technology, 14 April 1993, P/E 203.946/rev.

9 For a more detailed discussion of the content and scope of these basic rules, see, for example, Hancher EC Electricity Law, Chancery Law Publishing, 1992,

10 [1964] ECR 585.

11 For oil and petroleum products see, Case 78/83 Campus Oil Case C347/88 Commission v Greece [1990] ECR 4747; Case 266/81 SIOT [1983] ECR 731,Case 174/84 Bulk Oil v Sun [1986] ECR 539. For gas, see Cases 169/84 Cofaz I [1986] ECR 391 and Case C169/84 Cofaz II [1992] CMLR 177, and Cases 67,68 and 76/84 Gebroeders van der Kooy [1988] ECR 219.

12 Case T39/90 SEP v Commission;T-16/91 Rendo v Commission judgment of 17 November 1992; an appeal is pending in both cases.

13 On the potential conflict between the doctrine of supremacy and that of subsidiarity, see Cass, "*Subsidiarity - the word that saves Maastricht*", (1992) 29 CMLRev 1129 at 1131.

14 Decision 91/329 of 30 April 1991, O.J. 1991 L178/31. In its subsequent Decision relating to the privatization agreements in Northern Ireland, however, the Commission concluded that given the isolated position of the province, there was no inter-state trade effect - O.J. 1992 C92/5.

15 Council Directive 90/547, O.J. 1990 L313/30.

16 Council Directive 91/269, O.J. 1991 L147/37.

17 Council Directive 90/377, O.J. 1990 L185/16.

18 O.J. 1990 L297/1. This Directive will now be replaced by the so-called Consolidated Directive 93/37, O.J. 1993 L199.

19 For a more detailed discussion of these various measures as they affect the electricity markets, see my EC Electricity Law, op. cit.

20 On this subject, see generally, P-A. Trepte, Public Procurement in the EC, CCH 1992.

21 Titles III and IV of the Treaty of European Union amending the ECSC and Euratom Treaties do not provide for the insertion of a provision similar to Article 3b.

22 As far as environment is concerned, Article 130R will now be dropped, and Article 3b will apply to environmental policy also.

23 See generally, M. Wilke and H. Wallace, Subsidiarity: Approaches to Power Sharing in the Community, London, 1990: A. Adonis and A. Tyrie, Subsidiarity, IEA, London, 1990.

24 Annex 1 to Part A of the Conclusions of the Edinburgh European Council, Agence Europe no. 5878 bis.

25 N. Emilou, ***Subsidiarity, an effective barrier against the enterprise of ambition***, (1992) 17 ELRev 383.

26 See Commission of the EC, DGXVII, Energy in Europe. Annual Energy Review, 1993.

27 Council Directive 91/414, O.J. 1991 L242/1.

28 See further Council Regulation 1210/90 on the establishment of the European Environmental Agency and the European environmental information and observation network, O.J. 1990 L120/1.

29 The content of the two Draft Directives is analyzed extensively in other contributions to this volume:

30 O.J. 1992 C139/12.

31 Draft Report, PE 203.946, para., 8.

32 O.J. 1993, C73/31.

33 European Report, nr. 1846, 24 March 1993.

34 On the question of the justiciability of Article 3B and the problems of applying the tests of "necessity" and "effectiveness" which it appears to contain are discussed in Mckenzie Stuart, ***A Busted Flush*** in Curtin and O'Keefe (eds.), Essays in Honour of Justice O'Higgins, 1992. See also, N. Emilou, loc. cit. at n.25.

35 J. Vickers, *Government Regulation Policy*, (1991)7 Oxford Review of Economic Policy, 13-30.

36 For a short overview of this policy, see J. Naftel, *The Natural Death of a Natural Monopoly* 1993, 14 ECLR 105.

37 Directive 88/301 O.J. 1988 L131/73 .

38 Directive 90/388, O.J. 1990 L192/10.

39 Cases C202/88 France v Commission, judgment of March 19, 1991, [1991] ECR I-1223, and Cases C271. C281 and C289 /91 Spain, Italy and Belgium v Commission, judgment of 17 November 1992, nyr.

40 One issue which might deserve further consideration is whether the Commission may base measures requiring the withdrawal of exclusive rights to import or export or otherwise transport and supply energy on

Article 100A. Is it not constrained by the wording of the Treaty to use Article 90(3) if it wishes to achieve these goals?
Article 90(3) imposes a duty on the Commission to ensure that the provisions of Article 90(1) and (2) are applied. This could be interpreted to mean that the Commission must base its action on this Article. The Court of Justice has never been required to rule directly on this point. In its case law on Article 90(3) it has stated that the fact that the Council is also competent to adopt certain measures, for example, in the field of state aids, does not preclude the Commission from adopting similar measures to deal with problems expressly pertaining to the types of undertakings falling within the scope of Article 90(1)See Joined Cases 188-90/80 France Italy and the United Kingdom v Commission [1982] ECR 2545. See also Case C202/88, France v Commission, loc. cit.

41 Commissioner Van Miert's Speech of July 15, 1993, was reported in EC Energy Monthly, August 1993.

42 The Commission has submitted a separate proposal to the Council on access to hydrocarbon exploration and production, Com (92)100 final.

43 111 See Bell and Porter, *A Single European Market for Natural Gas*, [1991] 10 OGLTR 307.

44 Indeed in certain cases Community law requires it (see for example, Article 37 Euratom, and the interpretation of the scope of that Article in the Court's ruling in Case 187/87 Cattenom).

45 See further, B. Tenenbaum, R. Lock and J. Barker, *Electricity Privatization: Structural, Competitive and Regulatory Options*, Energy Policy, September 1992.

46 Recital 7 of the Amended proposal, not yet published.

CHAPTER 14

International law and energy regulation: The climate change convention

Philippe Sands[1]

Introduction

International protection of the atmosphere is a relatively recent subject for international legal regulation. With the exception of the EEC Regulations adopted in the late 1950's on motor vehicle emissions and the 1963 Test Ban Treaty, which was limited to atmosphere nuclear tests, and agreements on outer space, until 1979 no treaty sought, as its primary purpose, to place limits on the general right of states to permit the emission into the atmosphere of substances which were known or thought to cause pollution or environmental damage, including those from energy production. The issue of atmospheric pollution had, however, received international attention as early as the 1930s in the dispute between the United States and Canada over sulphur pollution from the Trail Smelter, which led to the important arbitral awards of 1938 and 1941. Recent treaties and other international acts have responded to specific threats to the quality of the atmosphere, including transboundary pollution by sulphur dioxide, nitrogen oxide and volatile organic compounds; the protection of the ozone layer; and the prevention of climate change. Of these, sulphur dioxide and greenhouse gas emissions are most closely tied to the energy sector.

Sulphur dioxide, which is caused by the combustion of high sulphur content fossil fuels (coal and oil), contributes to acid rain and is harmful to human health as a potent respiratory tract irritant. Combustion of fossil fuels, particularly from motor vehicles and power stations, also produces two oxides of nitrogen nitric oxide (NO) and nitrogen dioxide (NO_2), which are collectively known as NO_x. Ambient concentrations of NO_2 are generally considered to be too low to pose a significant threat to human health, but NO_x, together with hydrocarbons, are important precursors to the formation of tropospheric O_z and other photochemical oxidants. Sulphur and nitrogen oxides are transported by prevailing winds for distances up to 1,000 km from their original source before returning to the surface of the earth as wet or dry deposits. World-wide monitoring has established that North America and Europe receive fluxes of sulphur and nitrogen up to ten times the estimated natural flux. Adverse effects on the environment flowing from these deposits include acidification of fresh waters and terrestrial ecosystems. In the past 50 years the soils of many European forests have become 5 to 10 times more acidic.[2] Although these problems have so far been limited to developed countries, there are indications that certain tropical regions in developing countries, including southern China, south-western India, south-eastern Brazil and northern Venezuela may experience future problems with acidification, in large part due to rapid industrialization and the fact that tropical soils, already highly weathered and fairly acidic, are predisposed to acid damage.

International legal efforts to reduce emissions of sulphur dioxide and NOx emissions have led to a number of international agreements. These include:

- the United Nations Economic Commission for Europe (UNECE) Regulations Concerning Gaseous Pollutant Emissions from Motor Vehicles were adopted under the 1958 Agreement Concerning the Adoption of Uniform Conditions of Approval and Reciprocal Recognition of Approval for Motor Vehicle Equipment and Parts;[3]
- the 1979 UN/ECE Convention on Long-Range Transboundary Air Pollution (1979 Long-Range Transboundary Air Pollution (LRTAP) Convention),[4] together with protocols on the Reduction of Sulphur Emissions or Their Transboundary Fluxes by at Least 30 Per Cent (1985);[5] on the Control of Emissions of Nitrogen Oxides or Their Transboundary Fluxes (1988);[6] and the Control of Emissions of

Volatile Organic Compounds and Their Transboundary Fluxes' (1991);[7]

- the 1988 EC Large Combustion Directive, which allocates anthropogenic emission rights to each of the twelve Member states in respect of sulphur dioxide, nitrogen oxides and dust particles from large combustion plants;[8] and
- the 1991 Agreement Between the Government of the United States of America and the Government of Canada on Air Quality.[9]

These agreements, together with the 1985 Convention for the Protection of the Ozone Layer[10] and the 1987 Protocol on Substances that Deplete the Ozone Layer,[11] provided models for the elaboration of the most important international legal instrument to regulate energy use: the United Nations Framework Convention on Climate Change, which was opened for signature in June 1992 and entered into force in March 1994.

Climate change[12]

Introduction

The earth's climate is determined in large part by the presence in the atmosphere of naturally occurring greenhouse gases, including in particular water vapour, carbon dioxide, (CO_2), methane (CH_4), chlorofluorocarbons (CFCs), nitrous oxide (N_2O) and tropospheric ozone (O_z). These are transparent to incoming shortwave solar radiation but absorb and trap longwave radiation emitted by the earth's surface. Their presence exerts a warming influence on the earth. Scientific evidence suggests that continued increases in atmospheric concentrations of selected greenhouse gases due to human activities will lead to an enhanced 'greenhouse effect' and global climatic change.[13] Carbon dioxide emissions from the combustion of fossil fuels, production of cement, and agricultural and other land use, (including deforestation) is widely considered to be the most significant contribution to the threat of climate change, but global emissions of CFC-11 and 12, methane and nitrous oxide also pose a significant threat. In 1988 United Nations Environmental Programme (UNEP) and the World Meterological Organisation (WMO) established the Intergovernmental Panel on Climate Change (IPCC) to provide the international community with the scientific guidance necessary to take further action. The First Report of the IPCC, published in August 1990, predicted that on a 'business-as-usual' emissions scenario, the global mean temperatures could rise by an average rate of about 0.3°C per decade (with an uncertainty range of 0.2 - 0.5°C) during the next century with an uncertainty range of 30-110cm.[14] This could lead to an increase in global mean temperature of about 2°C above that occurring in pre-industrial times by the year 2025, and about 4°C by 2100. Such a rate of increase would be expected to lead to increased global average of rainfall by a few per cent by 2030, to diminution of areas of sea ice and snow cover, and to a rise in global mean sea level of 20cm by 2030 and 65 cm by the end of the next century.[15] In February 1992, the IPCC produced its Second Report, which concluded that findings of scientific research since 1990 did not affect the Working Group's understanding of the science of the greenhouse effect and either confirmed or did not justify altering the major conclusions of the 1990 Assessment.[16] The environmental, economic and social consequences of such rates of warming are described by the IPCC Working Group on Impacts.

The energy sector is the most important single source of greenhouse gases (GHGs), accounting for nearly 57% of radiative forcing from anthropogenic sources in the 1980s. Global energy and production has increased by roughly 50% over the past two decades, with fossil fuels (coal, oil and gas) together accounting for over 90% of that production. However, it plays a vitally important role in the economic development of all nations. The Energy and Industry subgroup of Working Group III of the IPCC (the Intergovernmental Panel on Climate Change, which was established in November 1988 in Geneva under the auspices of UNEP and WMO) recognised this fact in its report, which stated that their goal was to identify the paths to ensure that greenhouse gas emissions were compatible with the concept of sustainable development.

It is difficult to compare the relative contributions of greenhouse gases to radiative forcing due to the significant differences in the 'average atmospheric lifetimes of radiatively important gases'.[17] Models depend on a range of critical assumptions, the detailed discussion of which is beyond the scope of this chapter. The IPCC's Energy and Industry Sub-Group (EIS) was charged with, inter alia, defining policy options for national, regional and international responses to the possibility of climate change from GHG emissions produced by energy production, conversion and use.[18]

Between the years 1950 and 1970, fossil fuel generated energy use grew by around 4% per year in Western European countries. As a result of the OPEC oil price hikes in the 1970s and the corresponding slow down in economic growth, this figure fell between 1970 and 1985 to 0.4% per annum. This resulted also from structural changes in the OECD

(Organisation for Economic Co-operation Development) countries' energy sectors, primarily as a result of increased energy efficiency and reduced dependence on and use of imported oil.[19] Within the planned economies of Eastern Europe the annual rise in fossil fuel energy use was 5.2% between 1950 and 1985. Asian centrally planned economies showed a dramatic rise of 9.8% per annum over the same period, and the developing countries energy consumption grew at a rate of 5.7% in that period.[20]

The largest single source of GHG emission is carbon dioxide from the combustion of fossil fuels and other industrial sources.[21] The EIS forecast, in the absence of further policy and legislative measures, average annual growth of carbon dioxide emissions of 3.6% in the developing world and 0.8% in OECD countries. The share of carbon dioxide emission from the OECD countries would fall from 48% in 1985 to 33% by 2025, and Eastern European emissions would fall from 26% to 24% over the same period. By contrast, developing country emissions would rise from 26% of the global total in 1985 to 44% in 2025.[22]

In their assessment of these projections, the EIS identified the following as broad determinants of energy use which affect GHG emissions:

(a) population levels - those countries with rapid population growth rates are likely to experience a significant rise in energy use, with developing countries expected to account for more than 50% of the global increase in carbon dioxide emissions by 2025;[23]

(b) the level and structure of economic activity as measured by gross domestic product (developing countries are expected to see the demand for energy triple between 1985 and 2025);[24] and

(c) the carbon intensity of energy facilities.

The negotiation of a treaty to address climate change and its effects was formally set in motion by the UN General Assembly and its Specialized Agencies. In 1988, and again in 1989, the UN General Assembly determined that 'climate change is a common concern of mankind' and urged governments, intergovernmental and non-governmental organizations to collaborate in a concerted effort to prepare, as a matter of urgency, a framework convention on climate change.[25] The political process leading to the negotiation of a legal instrument was given further impetus by the 1990 Ministerial Declaration of the Second World Climate Conference,[26] which called for negotiations on an effective framework convention on climate change containing appropriate commitments to begin 'without delay'. In December 1990, the UN General Assembly established a single intergovernmental negotiating process under the auspices of the General

Assembly, supported by UNEP and WMO, for the preparation by an Intergovernmental Negotiating Committee for a framework Convention on Climate Change (INC/FCCC) of 'an effective framework convention on climate change, containing appropriate commitments, and any related instrument as might be agreed upon.'[27] The INC/FCCC held five sessions between February 1991 and May 1992 and the Convention was adopted at the close of the resumed fifth session, on 9th May 1992. The United Nations Framework Convention on Climate Change (1992 Climate Change Convention) will enter into force 90 days after the fiftieth ratification.[28] The Convention, which was signed by 155 states and the EC in June 1992 at United Nations Conference on Environment and Development (UNCED), is a package which contains elements for almost all the negotiating states but left none entirely satisfied. It reflected a compromise between those states which were seeking specific targets and timetables for emissions reductions, and those which wanted only a 'bare-bone' skeleton Convention which could serve as the basis for future Protocols, rather like the 1985 Vienna Convention.

Whatever the legal or substantive merits or demerits of the Convention, it went significantly beyond the scope of the 1985 Vienna Convention, which took nearly three times as long to negotiate amongst a much smaller group of states. The word 'Framework' in the title is something of a misnomer, since the 1992 Climate Change Convention established:

(a) commitments to stabilize greenhouse gas concentrations in the atmosphere at a safe level, over the long term, and to limit emissions of greenhouse gases by developed countries in accordance with soft targets and timetables;
(b) a financial mechanism and a commitment by certain developed country Parties to provide financial resources for meeting certain incremental costs and adaptation measures;
(c) two subsidiary bodies to the Conference of the Parties;
(d) a number of important guiding 'Principles'; and
(e) potentially innovative implementation and dispute settlement mechanisms.

Taken as a whole, the Convention marked further phase in the development of international environmental law. It was the first international environmental agreement to be negotiated by virtually the whole of the international community, with 143 states participating in the final session of the INC/FCCC. It is also an instrument which is potentially unique in the scope of its direct and indirect consequences: it is difficult to identify any type of human activity which will, over time, fall outside its scope. Affecting the vital economic interests of almost all states, it attempted to adopt a comprehensive approach to integrating environmental considerations into traditional economic development and defined, in legal terms, rights and obligations of different members of the international community in the quest for 'sustainable development' and the protection of the global climate.

The relationship between the Climate Change Convention and vital national economic, social and environmental interests was evident from the different interest groups of states which emerged during the negotiation. On the key issues, the Commitments section, this was not simply a North-South negotiation, as was clear from the failure of both the OECD and Group of 77 countries to reach common positions. Developed countries were far from united, with the United States alone in publicly opposing the adoption of specific targets and timetables and seeking to ensure a 'comprehensive' approach which deals with all greenhouse gas emissions, not just carbon dioxide. The economic implications of the Convention played a significant role in defining country positions, with Germany and Japan at the forefront of those developed countries viewing the Convention also as an instrument for gaining longer term competitive advantage by requiring the further development, production, use and dissemination of innovative new technologies. The differing economic capacities of developed countries, and in particular the problems faced by the former socialist countries of central and eastern Europe, led to a novel distinction being drawn in the Convention: for the purposes of differentiating those specific commitments relating to sources and sinks,[29] and those relating to finance, a distinction was drawn between all developed country Parties and developed Parties (included in Annex I)[30] and those developed country Parties and developed Parties not 'undergoing the process of transition to a market economy' (listed in Annex II).[31]

Developing countries were also divided. The oil producing countries, led by Saudi Arabia, strongly opposed any substantive obligations in the Convention, and clearly would not have been unhappy to see the negotiations fail altogether. The large industrializing developing countries, such as India and China, were understandably concerned to ensure that their economic development, including use of large coal reserves, should not in any way be limited. Developing countries with extensive forests, such as Brazil and Malaysia, were concerned to ensure that the primary emphasis of the Convention should be on limiting developed country emissions and not on protecting or enhancing developing countries' sinks (forests). And developing countries particularly vulnerable to the effects of climate change, such as the 37 member Alliance of Small Island states (AOSIS), sought a Convention with strong and enforceable commitments and an emphasis on the adverse effects of climate change. The fragmentation of countries into special interest groups created alliances between unlikely partners. On specific commitments relating to emissions the interests of the US, Saudi Arabia and China were broadly similar, as were those of the EEC and EFTA countries and AOSIS and certain African countries suffering from drought and desertification at the other end of the spectrum. It is in the context of these complex economic and environmental interests that the emergence of a package Convention must be understood.

Preamble, definition, objective and principles

The Convention's Preamble of 24 paragraphs is extensive enough to represent a wide range of interests. It contains a number of matters jettisoned from the 'Principles' article which are of some importance but which will have, in the context of this Convention, a less influential legal status as preambular paragraphs. The Preamble expressly recognizes, inter alia, 'the principle of sovereignty', that the largest share of historical and current global emissions has originated in developed countries, and includes (for the first time in a treaty) Principle 2 of the Rio de Janeiro Declaration on Environment and Development (rather than Principle 21 of the Stockholm Declaration). The Preamble also refers to the concepts of 'per capita emissions' and 'energy efficiency' which did not obtain sufficient support to be included in the operational part of the Convention.

Of note in the Definitions article is the omission of the concept of 'net emissions' (sources minus sinks, but no agreement was possible on whether to include natural sinks such as oceans), and the appearance of a bizarre footnote to the first title ('Definitions', Article 1) which states that 'Titles of articles are included solely to assist the reader'.[32]

The ultimate objective of the Climate Change Convention is to stabilize greenhouse gas concentrations in the atmosphere 'at a level that would prevent dangerous anthropogenic interference with the climate system', emphasizing that prevention of climate change is the primary objective.[33] However, the Convention implicitly recognizes that some climate change is inevitable, since the objective is to be achieved in such a way as to allow 'ecosystems to adapt naturally to climate change, to ensure that food production is not threatened and to enable economic development to proceed in a sustainable manner'.[34] Moreover, the text of the

Convention is liberally sprinkled with references to the 'effects' and 'adverse effects' of climate change (22 times), as well as references to 'vulnerability' and 'impacts' (seven times), supporting the view that the Convention also has the important, but unstated, objective of establishing an instrument to address the adverse effects of climate change and ensure that countries, particularly those most vulnerable, are able to prepare adequately for adaptation to the adverse effects of climate change.[35]

Article 3 of the Convention sets out a number of 'Principles' to guide the Parties in achieving the objective and implementing the provisions. Some of these mark new developments in international law and will provide an aid to interpreting some of the more ambiguous parts of the Convention, particularly relating to the commitments of developed country Parties. The obligation of Parties to protect the climate system should be 'on the basis of equity' and 'in accordance with their common but differentiated responsibilities and respective capabilities', in accordance with which developed country Parties should take the lead.[36] Parties should take measures, and adopt policies, which are 'precautionary', 'cost-effective', 'comprehensive', and which should take into account different 'socio-economic contexts'.[37] Climate change policies should also be integrated with national development programmes, and measures to combat climate change 'should not constitute a means of arbitrary or unjustifiable discrimination or a disguised restriction on international trade.'[38] This final principle is likely to be of some significance in relation to future trade disputes which may follow the adoption by a state or states of unilateral climate change-related trade prohibitions since it leaves open the possibility that certain such unilateral measures could be compatible with the Convention. Finally, throughout the Principles section, and elsewhere in the Convention, reference is made to the need to ensure 'sustainable economic growth' in order to address the problems of climate change.

General commitments

To achieve the objective all Parties are committed under Article 4(1) to take certain measures, taking into account their common but differentiated responsibilities and priorities, objectives and circumstances. These general commitments include, in particular, the development of national inventories of anthropogenic emissions by sources and removals by sinks of all greenhouse gases not controlled by the Montreal Protocol,[39] and the formulation and implementation of national and, where appropriate, regional programmes containing measures to mitigate climate change by addressing emissions and removals of these gases and be facilitation of adequate adaptation to climate change.[40] All Parties are required to promote and cooperate in the diffusion of technologies, practices and processes that control, reduce or prevent anthropogenic emissions of greenhouse gases not controlled by the Montreal Protocol, promote sustainable management, conservation and enhancement of sinks and reservoirs of these greenhouse gases, and co-operate in preparing for adaptation to the impacts of climate change.[41] All Parties are also required to take climate change into account, to the extent feasible, in their social, economic and environmental policies; to promote and co-operate in research, systematic observation and development of data archives to further understanding of climate change and response strategies; promote and co-operate in full, open, and prompt exchange of relevant information, and promote and co-operate in education, training and public awareness.[42]

Reporting

The Convention will establish broad reporting requirements for the communication of certain information, with specific provision for financial resources to be made available to developed country Parties. All Parties will be required to communicate to the Conference of the Parties information related to implementation, in accordance with Article 12, the communication of a national inventory of anthropogenic emissions by sources and removals by sinks of all greenhouse gases not controlled by the Montreal Protocol, a general description of steps taken or envisaged to implement the Convention; and any other relevant information including, if possible, material relevant for calculating global emission trends.[43] The effective implementation by developing country Parties of their communication commitments is linked to the effective implementation by developed country Parties of their financial commitments, including the need for adequacy and predictability in the flow of funds.[44] Annex I Parties are to include information relating to measures and policies to fulfil commitments under Article 4(2)(a) and (b), and a specific estimate of the effects those policies and measures will have on emissions and removals by the year 2000.[45] Annex II Parties shall include details of measures taken in accordance with Articles 4(3),(4) and (5).[46]

Initial communications for each Annex I Party must be made within 6 months of the entry into force for it of the Convention, and for each other Party within 3 years of entry into force for it, or upon the availability of financial resources under Article 4(3); least developed country Parties may make their initial communications at their discretion, and the timetable for subsequent communications is to be set by the Conference of the Parties.[47] Article 12 also provides for joint communication by a group of Parties, the protection of confidential information, and making communications public.[48]

Commitments: Sources and sinks

At the heart of the Convention are the specific commitments relating to sources and sinks of greenhouse gases binding on all developed country Parties and the EC under Article 4(2). The precise nature and extent of these commitments is unclear as a result of the convoluted language agreed to by way of compromise between various OECD members, and the different interests in and between developed and developing countries. The relevant provisions of Article 4(2), which stands as possibly the most opaque treaty language which has ever been drafted, provide:

(a) Each of these Parties shall adopt national policies and take corresponding measures on the mitigation of climate change, by limiting its anthropogenic emissions of greenhouse gases and protecting and enhancing its greenhouse gas sinks and reservoirs. These policies and measures will demonstrate that developed countries are taking the lead in modifying longer-term trends in anthropogenic emissions consistent with the objective of this Convention, recognizing that the return by the end of the present decade to earlier levels of anthropogenic emissions of carbon dioxide and other greenhouse gases not controlled by the Montreal Protocol would contribute to such modification, and taking into account the differences in these Parties' starting points and approaches, economic structures and resource bases, the need to maintain strong and sustainable economic growth, available technologies and other individual circumstances, as well as the need for equitable and appropriate contributions by each of these Parties to the global effort regarding that objective. These Parties may implement such policies and measures jointly with other Parties and may assist other Parties in contributing to the achievement of the Convention and, in particular, that of this sub-paragraph;

(b) In order to promote progress to this end, each of these Parties shall communicate, within six months of the entry into force of the Convention for it and periodically thereafter, and in accordance with article 12, detailed information on its policies and measures referred to in subparagraph (a) above, as well as on its resulting projected anthropogenic emissions by sources and removals by sinks of greenhouse gases not controlled by the Montreal Protocol for the period referred to in subparagraph (a), with the aim of returning individually or jointly to their 1990 levels of these anthropogenic emissions of carbon dioxide and other greenhouse gases not controlled by the Montreal Protocol ...'

Whether these two paragraphs be interpreted to amount to something meaningful and capable of implementation is open to question. They do not reflect a commitment to stabilise carbon dioxide and other greenhouse gas emissions by the year 2000 at 1990 levels. Even when read together the two paragraphs do not amount to a clear commitment to return emissions of carbon dioxide and other greenhouse gases not controlled by the Montreal Protocol to 1990 levels by the year 2000 (and with Parties retaining the right to increase their emissions thereafter). Article 4(2)(a) requires only the 'limitation' by each developed country Party of its anthropogenic emissions of greenhouse gases, as opposed to stabilization or reduction. It also recognises, in an especially unattractive 117 word sentence, that the return to 'earlier levels' by the year 2000 'would' contribute to the modification of longer term trends in emissions consistent with the objective of the Convention. This is clearly something other than a provision requiring a mandatory return to a specified earlier level by a specified date. Also noteworthy is the absence of a commitment to keep emissions no higher than 1990 levels after 2000. Further, each Party's contribution is dependent on a series of factors, including its economic structure, resource base, starting point, and approach, as well as the application of 'equity'. Article 4(2)(b) is perhaps a little less opaque. It requires information to be provided on projected anthropogenic emissions for the period up to 2000, and establishes only the 'aim' of returning to 1990 levels without providing a date by when such a return should be achieved. The most that can reasonably be said of these provisions is that they establish soft targets and timetables with a large number of loopholes; the adequacy of Articles 4(2)(a) and (b) will be reviewed at the first session of the Conference of the Parties and regularly thereafter, with a second review to take place before 31 December 1998.[49]

The Convention goes further than the 1987 Montreal Protocol by allowing for possible 'joint implementation' by Annex I Parties of their policies and measures, subject to further decisions to be taken by the

Conference of the Parties regarding criteria for such 'joint implementation'.[50] This will provide the foundation for the efforts of those states which sought to ensure that emissions reductions should be carried out in the most 'cost-effective' way possible. The Convention additionally requires that 'a certain degree of flexibility' should be allowed to developed country parties 'undergoing the process of transition to a market economy'.[51] Parties are also to take into consideration in the implementation of commitments the situation of Parties, particularly developing country Parties, with economies vulnerable to the adverse effects of implementation of response measures.[52]

The calculation of emissions by sources and removal by sinks will take into account the best available scientific knowledge, pending agreement by the Conference of the Parties on common methodologies.[53] Each developed country Party is also required to co-ordinate relevant economic and administrative instruments and identify and periodically review its own policies and practices which encourage activities that lead to greater levels of anthropogenic emissions.[54]

Commitments: financial resources and technology transfer

Annex II Parties undertake specific financial commitments. They agree to provide 'new and additional' financial resources to meet the 'agreed full costs' incurred by developing country Parties in fulfilling their commitment to communicate information relating to implementation (Article 12), and to provide such financial resources needed by developing country Parties 'to meet the agreed full incremental costs of implementing measures' relating to their general commitments under Article 4(1) and which are agreed between the developing country Party and the entity responsible for the financial mechanism.[55] Annex II Parties also undertake to assist developing country Parties that are 'particularly vulnerable to the adverse effects of climate change in meeting costs of adaptation to those adverse effects.'[56] In what amounts to an implicit acceptance by developed country Parties of responsibility for causing climate change, Article 4(4) may ultimately emerge as one of the more unusual, contentious and perhaps costly, commitments in the Convention.

In the implementation of Article 4 the Parties are required to give full consideration to the actions necessary to meet specific needs and concerns of developing country Parties arising from the adverse effects of climate change, and/or the impact of implementing response measures, including actions related to funding, insurance and the transfer of technology, including least developed countries.[57] Certain categories of countries are identified, including small island countries, countries with low-lying coastal areas, countries with areas liable to drought and desertification, and countries whose economies are highly dependent on income generated from, or the consumption of, fossil fuels.

Annex II Parties are required to take all practicable steps to promote, facilitate and finance the transfer of, or access to, environmentally sound technologies and know-how, and support the development of endogenous capacities and technologies of developing country Parties.[58] In the short term the financial mechanism is likely to devote a significant proportion of the available financial resources to technology transfer.

Institutions and arrangement

The Climate Change Convention establishes a Conference of the Parties, a Secretariat, two subsidiary bodies, and a financial mechanism.[59] The Conference of the Parties is the supreme body of the Convention, entrusted with keeping the implementation of the Convention under regular review and making decisions to promote its effective implementation.[60] It will meet for the first time within a year of the entry into force of the Convention, and thereafter every year unless otherwise decided by the Conference of the Parties.[61] It has a large number of specifically enumerated functions, including:

- to periodically examine the obligations of the Parties;
- to facilitate the co-ordination of measures;
- to promote and guide comparable methodologies for preparing inventories of greenhouse gas emissions;
- to assess the implementation of the Convention by all Parties and the overall effect of measures; and
- to adopt regular reports on the implementation of the Convention.

A multidisciplinary Subsidiary Body for Scientific and Technological Advice will be established to provide information on scientific and technological matters to the Conference of the Parties.[62] A Subsidiary Body for Implementation will be established to assist the Conference of the Parties in the assessment and review of the implementation of the Convention,[63] and this is could emerge as one of the more innovative aspects of the Convention. Although some states wanted to limit participation, both subsidiary bodies will be open to participation by all Parties.

The Convention defines a financial mechanism for the provision of financial resources on a grant or concessional basis, including for the transfer of technology.[64] After specific commitments this was the most disputed aspect of the Convention. The mechanism will function under the guidance of, and be accountable to, the Conference of the Parties, which will be responsible for its policies, programme priorities and eligibility criteria, and its operation is to be entrusted to one or more existing international entities.[65] The mechanism is to have an equitable and balanced representation of all Parties within a transparent system of governance. No permanent mechanism has been designated. The Global Environment Facility (GEF) of the UNDP, UNEP and IBRD is entrusted with the operation of the financial mechanism on an interim basis, and for that purpose it is to be 'appropriately restructured and its membership made universal' to enable it to fulfil the requirements of Article 11.

Implementation and dispute settlement

Apart from the role of the Conference of the Parties and the Subsidiary Body for Implementation, the Convention provides for the possibility of establishing a 'multilateral consultative process' for the resolution of implementation questions, which will be available to Parties on their request.[66] This whittles down two more ambitious original proposals. Additionally, a dispute settlement Article provides for possible compulsory recourse to arbitration or the International Court of Justice with the consent of the relevant parties to a dispute, as well as the possibility for the compulsory establishment of a conciliation commission, with the power to make a recommendatory award, at the request of one of the parties to a dispute twelve months after notification of the dispute.[67] The Convention provides for amendment, the adoption and amendment of Annexes, and the adoption of Protocols.[68] No reservations are permitted.[69] Prior to its entry into force, Article 21 of the Conventions establishes interim arrangements concerning the designation of an interim secretariat, co-operation with the IPCC and other scientific bodies[70], and designation of the GEF as the interim financial mechanism.

Agenda 21

Agenda 21 is a non-binding blueprint and action plan for a 'global partnership for sustainable development'.[71] It runs to forty chapters and was negotiated over a two year period. Chapter 9 of Agenda 21, on 'Protection of the atmosphere', was among the most controversial subjects addressed at UNCED, giving evidence of the potential impacts of international environmental regulation on the fundamental economic interests of many states, in particular energy regulation. A number of OPEC states, led by Saudi Arabia, Libya and Kuwait opposed the Chapter in its entirety. The political sensitivity of Chapter 9 is clear from its introduction, which states that 'the recommendations contained in this chapter do not oblige any Government to take measures which exceed the provisions of the 1985 Vienna Convention, the 1987 Montreal Protocol as amended, the 1992 Climate Change Convention, and any other international, including regional, instruments.[72] On the other hand, and in order to achieve a balance and with an eye to possible future trade disputes over unilateral national atmospheric protection and energy standards, it is also stated that 'within the framework of this chapter, Governments are free to carry out additional measures which are consistent with those legal instruments'.[73] The Chapter establishes four programme areas which emphasize the priorities of the international community in coming years, including programmes on promoting sustainable development, and transboundary air pollution.

Promoting sustainable development

This programme head contains sections on energy development efficiency and consumption, transportation, industrial development, and resource development and land use. With regard to the energy sector the programme recognises that:

> the need to control atmospheric emissions of greenhouse and other gases and substances will increasingly need to be based on efficiency in energy production, transmission, distribution and consumption, and on growing reliance on environmentally sound energy systems, particularly new and renewable sources of energy.[74]

The basic and ultimate objective of the programme area which could serve as the basis for the elaboration of a future global energy policy is:

> to reduce adverse effects on the atmosphere from the energy sector by promoting policies or programmes, as appropriate, to increase the contribution of environmentally safe and sound and cost-

> effective energy systems, particularly new and renewable ones, through less polluting and more efficient energy production, transmission, distribution and use. The objective should reflect the need for equity, adequate energy supplies and increasing energy consumption in developing countries

as well as the needs of certain vulnerable countries.[75] The programme area identifies activities to be carried out by Governments which could serve as a possible basis for future international legislation, and is important as the first occasion on which the whole of the international community has come together to propose the basis for future international energy policy. The programme area seeks to promote research, development, transfer and use of improved energy-efficient technologies and practices and [safe and] sound energy systems; promote the development of capacities to develop, produce and use increasingly efficient and less polluting forms of energy, review current energy supply mixes, coordinate energy plans regionally and sub-regionally, promote cost-effective policies or programmes (including administrative, social and economic measures) to improve energy efficiency, promote energy efficiency and emission standards at the national level, encourage education and public awareness about energy efficiency and environmentally sound energy systems, and establish energy efficiency labelling programmes.[76]

Apart from the rules established by the ICAO (International Civil Aviation Organisation) in respect of aircraft emissions, the UNECE rules for motor vehicles, and the rules of EC law, international legislation on pollution from the transport sector is minimal. In the UN system no single institution or body has responsibility for transport in general, and the regulation of its environmental protection standards in particular. The Agenda 21 programme area on transportation is therefore an indicator of possible future international legal developments. The overall objective of the programme area on transportation is to develop cost effective policies and programmes to limit, reduce or control harmful atmospheric emissions and other adverse environmental effects of the transport sector, taking into account development priorities, safety and national circumstances.[77] The programme area on industrial development seeks to encourage industrial development in ways that minimize adverse impacts on the atmosphere by increasing industry's efficiency in consumption and production, improving pollution-abatement technologies, and developing new environmentally sound technologies.[78] Both programme areas are to be achieved through measures taken by Governments, intergovernmental and non-governmental organizations and the private sector to, inter alia, develop cost-effective, more efficient and less-polluting transport systems, particularly rural and urban mass transit; to encourage the transfer of resource efficient and less polluting transport and other industrial technologies, particularly to developing countries; to develop technologies, products and processes which are less polluting and more efficient in their use of natural resources; promote administrative, social and economic measures to encourage transport modes and industrial practices which minimize adverse impacts on the atmosphere.[79]

The programme area on terrestrial and marine resource development and land use is designed to reduce atmospheric pollution and limit anthropogenic emissions of greenhouse gases, conserve, sustainably manage and enhance greenhouse gas sinks and natural and environmental resources; and ensure that atmospheric changes are fully taken into account in planning and implementing policies and programmes.[80]

Transboundary atmospheric pollution

Central to this programme area is the objective of establishing new regional agreements for limiting transboundary air pollution, based upon the experience of the 1979 LRTAP Convention and its Protocols,and ensuring the implementation of that agreement.[81] The programme also calls for the strengthening of systematic observation and monitoring, the development and exchange of emission control technologies from mobile and stationary sources, the establishment and strengthening of early warning systems and response mechanisms for transboundary air pollution from industrial and nuclear accidents and natural disasters.[82]

Conclusions

Developments at the regional and global level have taken place which now establish significant limitations on the right of states to allow emissions of gases which cause urban and transboundary air pollution and increased atmospheric concentrations of carbon dioxide and other greenhouse gases produced by fossil fuel combustion. In so doing, an array of regulatory techniques has been deployed, including the use of a 'target-and-timetable' approach, differentiated commitments for developed and developing countries, and a proposed carbon/energy tax designed to put in place incentives to limit fossil fuel use. As one commentator has noted, the 'convention on climate change has the potential to affect economic and social activities profoundly, much more so than other international environmental agreements dealing with problems ... which have more limited and easily addressed causes'.[83]

Taken in a relative context, these are impressive and relatively speedy developments. Nevertheless, major gaps remain to be addressed. First, in relation to urban and transboundary air pollution, the rules are almost entirely applicable to developed countries in the OECD/ECE/EC context; as rapid industrialisation takes place in other regions there is a need to develop rules to address these related problems, perhaps through the development of a regional approach modelled upon UNEP's Regional Seas Programme. Second, the 1992 Climate Change Convention should be brought into force promptly, and the financial arrangements necessary to encourage the participation of, in particular, the largest developing countries, should be assured. Third, greater attention needs to be given to the enforcement of these agreements, including in respect of meeting reporting requirements and providing independent verification that targets and timetables have been and are being complied with. And fourth, international lawyers will need to address a new range of legal issues thrown up by the development of innovative international mechanisms and techniques to assist in compliance: carbon taxes and 'joint implementation' are amongst the new techniques which raise both economic and legal questions which have not been fully addressed or understood and will require the further development of public international law.

Notes

1 Barrister; Lecturer in Law, School of Oriental and African Studies, London University; Legal Director, Foundation for International Environmental Law and Development; Visiting Professor, New York University. The author would like to thank Paul Osborne for his assistance in preparing this Chapter.

2 World Resources 1992-93, p. 198.

3 Geneva, 20 March 1958, in force 20 June 1959; 335 U.N.T.S. 211.

4 Geneva, 13 November 1979, in force 16 March 1983, 18 I.L.M. 1442 (1979).

5 Helsinki, 8 July 1985, in force 2 September 1987; 27 I.L.M. 707 (1988).

6 Sofia, 31 October 1988, in force 14 February 1991; 28 I.L.M. 214 (1989).

7 Geneva, 18 November 1991, not in force; 31 I.L.M. 568 (1992).

8 Council Directive 88/609/EEC of 24 November 1988 on the limitation of emissions of certain pollutants into the air from large combustion plants, as amended, OJ L 336, 7.12.1988, p. 1. A 'large' combustion plant is one with a rated thermal input equal to or greater than 50 MW: Art. 1.

9 Ottawa, 13 March 1991, in force 13 March 1991; 30 I.L.M.(1991).

10 Vienna, 22 March 1985, in force 22 September 1988, 26 ILM 1529 (1987).

11 Montreal, 16 September 1987, in force 1 January 1989, 26 ILM 154 (1987); amended and adjusted in 1990 and 1992; see R. Benedick, Ozone Diplomacy: New Directions in Safeguarding the Planet (1991)

12 See E. Barratt-Brown et al, 'A Forum for Action on Global Warming: the U.N. Framework Convention on Climate Change', 4 ColJIELP 103 (1993); D. Bodansky, The United Nations Framework Climate Change Convention: A Commentary, 18 YaleJIL 451 (1993); R. Churchill and D. Freestone (eds.), International Law and Global Climate Change (1991); M. Grubb, 'The Climate Change Convention: An Assessment' 15 IER 540; T. Iwama (ed.), Policies and Laws on Global Warming: International and Comparative Analysis (1991); P. Sands, 'The United Nations Framework Convention on Climate Change', 1 RECIEL 270 (1992); 'Selected International legal Materials on Global warming and Climate Change', 5 AmUJILP 513 (1990); C. Stone, 'The Global Warming Crisis, if There Is One, and the Law', 5 AmUJILP 497 (1990); D. Zaelke and J. Cameron, 'Global Warming and Climate Change: An Overview of international Legal Process', 5 AmUJILP 248 (1990).

13 See UNEP, Environmental Data Report, 3-9 and 121-130 (1991); the 1992 Climate Change Convention defines 'greenhouse gases' as 'those gaseous constituents of the atmosphere, both natural and anthropogenic, that absorb and re-emit infra-red radiation': Art. 1(5).

14 IPCC, Climate Change: the IPCC Scientific Assessment (1990).

15 The 'business-as-usual' scenario assumed a continued reliance on coal and oil, modest improvements in energy efficiency, limited controls on emissions of carbon dioxide, continued deforestation, uncontrolled emissions of methane and nitrous oxide from agricultural sources, and a reduction of CFCs in line with the 1987 Montreal Protocol. A rise in sea level also has social and economic consequences: wide scale population movements would cause unemployment, discrimination and disruption of traditional cultures. In resource terms, food production could be affected by "substantial dislocations in the agricultural sector". Biodiversity and

composition of species could be threatened, with the threat of extinction of fauna and competition for land with man. The uncertainty of crop yields due to shifting climatic zones could result in shortages and impact on employment and agricultural incomes.

16 IPCC, 1992 Supplement, (1992). Section II, para. 2.

17 WMO and UNEP, Climate Change - The IPCC Response Strategies (1991), 53.

18 Id., 49.

19 Id., 53.

20 World Resources Institute/UNEP/UNDP, World Resources 1992-93 (1992), 145.

21 Id., 210.

22 Supra. note 14, 58.

23 Supra. note 18, 149

24 World Resources 1992-3, p.149.

25 UN General Assembly resolution 43/53, 6 December 1988; UN General Assembly resolution 44/207, 22 December 1989.

26 U.N. Doc. A/45/696/Add.1, Annex III (1990).

27 GA Res. 45/221, 21 December 1990.

28 New York, 9 May 1992, not in force, 31 ILM 849 (1992); Art. 23(1). The Convention had attracted 26 ratifications within a year of its adoption and is expected to enter into force in 1994.

29 Under the Convention a 'source' is 'any process or activity which releases a greenhouse gas, a aerosol or a precursor of a greenhouse gas into the atmosphere: Art. 1(9); a 'sink' is 'any process, activity or mechanisms which removes a greenhouse gas or a precursor of a greenhouse gas from the atmosphere': Art. 1(8).

30 Annex I lists all the OECD countries and the EC (for which the term 'developed Party' was used, apparently for the first time in international law), as listed in Annex II, plus eleven former socialist countries: Belarus, Bulgaria, Czechoslovakia, Estonia, Hungary, Latvia, Lithuania, Poland, Rumania, Russia and Ukraine. Albania, Yugoslavia and certain members of the Commonwealth of Independent states appear in neither Annex and must therefore be deemed to be developing countries within the meaning of the Convention.

31 Annex II lists all OECD member countries (Australia, Austria, Belgium, Canada, Denmark, Finland, France, Germany, Greece, Iceland, Ireland, Italy, Japan, Luxembourg, Netherlands, New Zealand, Norway, Portugal,

Spain, Sweden, Switzerland, Turkey, United Kingdom, United States) and the EC. Turkey was apparently included in this list by oversight.

32 On the possible legal consequences of this footnotes see Chapter 6, note.

33 Art. 2. The 'climate system' is defined as 'the totality of the atmosphere, hydrosphere, biosphere and geosphere and their interactions': Art. 1(3); 'climate change' is 'a change of climate which is attributed directly or indirectly to human activity that alters the composition of the global atmosphere and which is in addition to natural climate variability observed over comparable time periods': Art. 1(2).

34 Id.

35 'Adverse effects of climate change' means 'change in the physical environment or biota resulting from climate change which have significant deleterious effects on the composition, resilience or productivity of natural and managed ecosystems or on the operation of socio-economic systems or on human health and welfare': Art. 1(1).

36 Art. 3(1).

37 Art. 3(3).

38 Art. 3(5).

39 Art. 4(1)(a).

40 Art. 4(1)(b).

41 Art. 4(1)(c)-(e); a 'reservoir' is defined as 'a component or components of the climate system where a greenhouse gas or a precursor of a greenhouse gas is stored': Art. 1(7).

42 Art. 4(1)(f)-(i).

43 Arts. 4(1)(j) and 12(1).

44 Art. 4(3) and (7).

45 Art. 12(2).

46 Art. 12(3).

47 Art. 12(5).

48 Art. 12(8)-(10).

49 Art. 4(2)(d).

50 Art. 4(2)(a) and (d).

51 Art. 4(6).

52 Art. 4(10).

53 Art. 4(2)(c).

54 Art. 4(2)(e).

55 Art. 4(3).

56 Art. 4(4).

57 Art. 4(8) and (9).

58 Art. 4(5).
59 Arts. 7 to 11.
60 Art. 7(2).
61 Art. 7(4).
62 Art. 9(1).
63 Art. 10(1).
64 Art. 11(1).
65 Art. 11(1)-(3).
66 Art. 13.
67 Art. 14.
68 Art. 15 to 17.
69 Art. 24.
70 Art. 21(1) and (2). The Secretariat of the INC, established by UN General Assembly Resolution 45/212, will act as interim secretariat until the completion of the first session of the Conference of the Parties. GA Res. 47/195 of 22 December 1992 decides that the INC will continue to function in order to prepare for the first Conference of the Parties.
71 Report of the United Nations Conference on Environment and Development, A/CONF.151/26/Rev.1 (Vol. 1) (1993).
72 Para. 9.2.
73 Id.
74 Para. 9.9. New and renewable sources of energy are defined as 'solar thermal, solar photovoltaic, wind, hydro, biomass, geothermal ocean, animal and human power, as referred to in the reports of the Committee on the Development and Utilization of New and Renewable Sources of Energy', prepared specifically for UNCED: see A/CONF.151/PC/119 and A/AC.218/1992/5.
75 Para. 9.11.
76 Para. 9.12.
77 Para. 9.14.
78 Para. 9.17.
79 Paras. 9.15 and 9.18.
80 Para. 9.20.
81 Paras. 9.26, 9.27(e), and 9.28(a).
82 Paras. 9.27 and 9.28.
83 D. Bodansky, 'The United Nations Framework Convention on Climate Change: A Commentary', 18 YaleJIL 451, 476 (1993).

CHAPTER 15

Electricity industry regulation and environmental issues: The view from OFFER

Peter Carter

Introduction

In this paper, I intend to discuss the interaction of environmental and economic regulation from a practical rather than a theoretical standpoint. I will describe how we in the Office of Electricity Regulation (OFFER) have attempted to deal with major environmental issues over the past three years. I will first outline the responsibilities of the Director General of Electricity Supply with respect to environmental matters and discuss the way in which he has attempted to discharge those responsibilities, with particular reference to energy efficiency and climate change. I will also discuss some of the possible implications of limits on sulphur dioxide emissions.

Statutory responsibilities

The Office of Electricity Regulation, and the post of Director General of Electricity Supply, have responsibilities regarding competition, regulation and consumer issues in the electricity industry in England and Wales, and in Scotland. The responsibilities of the Director are set out in the Electricity Act 1989. Section 3 of the Act says that he has a duty to exercise his functions under the Act in the manner which he considers is best calculated to achieve three objectives - securing that all reasonable demands for electricity are satisfied; securing that licence holders are able to finance the carrying on of their activities; and to promote competition in the generation and supply of electricity.

Section 3(3) lays on the Director some subsidiary duties. These include protecting the interests of electricity consumers, both as regards prices and as regards continuity and quality of supplies; and promoting efficiency and economy on the part of persons authorised by licences to supply or transmit electricity and the efficient use of electricity supplied to consumers. It also lays on him a duty to take into account, in exercising his functions, the effect on the physical environment of activities connected with the generation, transmission or supply of electricity.

Two other provisions of the 1989 Act concern environmental matters. Section 41 empowers the Director to determine Standards of Performance to be achieved by Public Electricity Suppliers in connection with promoting the efficient use of electricity by consumers.

Finally, Schedule 9 to the Act provides that a licence holder under the Act or someone authorised by exemption to generate or supply electricity should, in formulating proposals, for example, the construction of a new power station, have regard to the desirability of preserving natural beauty, of conserving flora, fauna and geological or physiographical features of special interest and of protecting sites, buildings and objects of architectural, historic or archaeological interest. The Schedule requires each licence holder to draw up a statement setting out the manner in which he proposes to perform his duty under the Schedule, including in particular the consultation procedures (e.g. with public bodies with responsibilities for the natural environment) which he intends to follow.

The Gas Act 1986 (as subsequently amended) contains provisions as to environmental duties of the gas regulator which are broadly similar, though not identical, to the provisions of the Electricity Act. The Director General of Gas Supply, like the Director General of Electricity Supply, has a subsidiary duty to promote efficiency and economy on the part of suppliers and consumers but does not, however, have a duty to take into account the effect on the physical environment of activities connected with the gas industry. The Gas Act did not originally give the Director General of Gas Supply power to draw up Standards of Performance for the industry with respect to promoting the efficient use of energy, though subsequent legislation, in 1992, provided such a power, in terms very close to Section 41 of the Electricity Act. The Gas Act contains no provision corresponding to Schedule 9 of the Electricity Act.

Comment

It is apparent then that the legislation has not drawn a tidy dividing line between the economic functions of the Director of Electricity Supply on one side and environmental regulation on the other, but has rather given the Director General certain duties with respect to environmental matters, particularly with respect to energy efficiency, which he needs to consider alongside his economic functions. But the legislation gives the Director General few powers with respect to environmental matters other than a power to set Standards of Performance for energy efficiency. This reflects the view that it is not for industry regulators to determine how tight environmental regulation should be. That task properly falls to others. Industry regulators may, however, consider whether their primary economic functions should be exercised in such a way as to promote desirable environmental objectives, or at least so as to avoid undesirable environmental consequences. It may also be appropriate for them to point out where necessary that environmental regulation has implications for costs and prices, and to quantify these where possible, and to point to areas where the mode of environmental regulation may cause unnecessary difficulties for, for example, competition and costs.

The Government's recent consultation paper on Climate Change and the UK national programme for carbon dioxide emissions pointed out that in 1990 power stations contributed 34% of UK carbon dioxide emissions by source. While recognising the potential impact of changes in the mix of fuels used for electricity generation, for example the move from coal to gas, and the contribution that could be made by combined heat and power (CHP), the document is clear that a major part of any national programme to stabilise or reduce carbon dioxide emissions must lie in a sustained effort to increase the efficiency with which energy is used.

This is one of the most important interfaces between the economic regulation of the energy industries and environmental policy objectives. It is OFFER's belief that measures which it might put in place to encourage energy efficiency need to meet four broad criteria:

- they must give the electricity companies real incentives to promote the efficient use of energy;

- they must be compatible with competition in electricity generation and supply, and with an efficient and profitable industry;
- they must protect the interests of customers and avoid undue discrimination between customers;
- they must not cause any undue distortions in competition in the energy supply industry.

Energy efficiency

In the light of these considerations, and of his duties under the Electricity Act, what has the Director General of Electricity Supply done to promote energy efficiency?

First, he has sought to gain fuller recognition for the role that competition can play in improving the efficiency of energy use. For example, competition in generation gives an incentive to generators to reduce the costs of generation. Fuel is by far the largest single component of a generator's cost. Promoting competition in generation is therefore a particularly important potential means of increasing energy efficiency.

Competition in supply also has a role in energy efficiency, and this is likely to become more important as more customers move into this increasingly competitive market. From 1 April 1994, all customers at premises with a maximum demand in excess of 100kw have been able to choose their own electricity supplier, and from 1 April 1998 all consumers will be able to do so.

Competition in supply can promote energy efficiency in several ways. An important issue in relation to energy efficiency is that consumers should receive accurate and consistent price signals about the real cost of resources which they are using. Competition in supply will limit or remove a company's ability to send misleading price signals to its customers, for example by under pricing peak electricity at the expense of off-peak electricity or other services. Recent advances in metering technology, which will facilitate the development of competition in electricity supply, will also enable companies to offer more price-reflective tariffs, which will give consumers the opportunity to save money by using electricity more efficiently.

Competition in supply gives suppliers stronger incentive to differentiate their services. Some suppliers might decide to do this by offering advice, assistance or energy management services to their customers.

Second, there is now a recognition that the structure of some of the original price controls on the non-competitive areas of the electricity industry (for example the controls on transmission and distribution, and on the part of the supply market which is not yet open to competition) may have given the electricity companies an artificial incentive to sell more electricity. Broadly, this effect can arise if the 'revenue driver' of the control gives significantly greater weight to the number of units of electricity sold than would be appropriate having regard to the underlying cost drivers of the business.

For example, in the new control on the transmission business of the National Grid Company, which took effect from 1 April 1993, we have sought to remove artificial incentives on the company to transmit more electricity. Similar considerations have been taken into account in the revisions of the controls on the transmission businesses of the two Scottish companies. The proposal for a new control on the supply business of the regional electricity companies (RECs) in England and Wales, which we published in July 1993 and which has now been implemented, considerably reduces the dependence of allowed revenue on the quantities of electricity sold, and thereby removes a bias towards selling more units. We will shortly complete a review of the price controls on the distribution businesses of the electricity companies both in England and Wales and in Scotland, and energy efficiency will be an issue in this review also.

There is considerable experience in the United States of regulatory intervention to improve the incentives for energy efficiency by electricity utilities. There are, of course, significant differences between the UK and US situations. In particular, the industry in the United States is characterised by vertical integration, local monopoly in supply, and a regulatory regime which controls prices so as to cover operating costs plus a return on approved investment. These differences mean that the US experience cannot be directly translated to the UK in all cases. Nonetheless OFFER recognises that the broad principle remains valid, that it is undesirable that the choice between demand side and supply side investments should be artificially distorted.

Third, we recognise that electricity suppliers are well placed to encourage more efficient use of electricity by their customers. In the competitive market, companies already have an incentive to encourage energy efficiency and this incentive will grow as the competitive market expands. In some industrial markets, electricity companies have won customers from gas by assisting in the development of processes which use electricity to high efficiency. As regards their domestic customers, who are not yet able to choose their electricity supplier, the licences of all 14 public electricity suppliers in Great Britain require them to draw up and make available to customers a Code of Practice containing advice on how to use electricity efficiently.

In connection with the new supply price control in England and Wales we have taken matters an important step forward by requiring the companies to explore more thoroughly the potential for cost effective demand side management measures. The new price control includes an annual allowance of £1 per customer in the franchise (i.e. non-competitive) market, which the companies are obliged to spend on projects designed to achieve improvements in the efficient use of energy for customers in the franchise market. New standards of performance have be drawn up to ensure that the money is spent as intended, and to ensure that the projects are properly monitored.

The published proposal for the new supply price control drew attention to some of the considerations which the Director had to bear in mind in coming to the conclusion that this was an appropriate way forward, and that £1 per franchise customer was an appropriate amount to allow.

There were, first of all, arguments that significantly more needed to be spent in order to meet the Government's commitment to stabilise carbon dioxide emissions. While we agreed that electricity customers should contribute to this aim, we concluded that they should not bear a disproportionate burden, particularly since domestic electricity customers would already be paying higher prices through the introduction of VAT on fuel from April 1994. We noted too that other factors - competition, changes in the generation fuel mix, changes in regulatory incentives - would reinforce the electricity industry's contribution to reducing carbon dioxide emissions. We also believed that it was important to bear in mind the redistributive effects of the expenditure, since although all franchise customers would pay, some would benefit to a greater degree than others, or sooner, from lower individual fuel bills. All in all, we concluded that if significantly more money needed to be raised from electricity consumers to fund energy efficiency programmes, then this raised issues more appropriately dealt with by Parliament through general fiscal policy rather than price control mechanisms proposed by an economic regulator.

The case of sulphur dioxide emissions

I turn now to the most important of the other environmental issues that affect the electricity industry, the problem of sulphur dioxide emissions from power stations. Although the Director's responsibilities with respect to this issue are only indirect, it does nonetheless have important implications for him in exercising his duties to promote competition and protect consumers.

The UK regime for control and reduction of sulphur dioxide emissions derives from the European Communities' Large Combustion Plant Directive. This requires governments to secure a planned reduction of 60% on the 1980 level of sulphur dioxide emissions by 2003. The Directive is implemented in the UK under the Environment Protection Act 1990, which empowers the government to make plans for establishing national emissions limits and for the allocation of quotas to parties responsible for such emissions. Under the plan, each generator has an overall emission quota, known as its 'bubble', which it must not exceed. Specific limits are set annually for each generating plant by HM Inspectorate of Pollution (or, in Scotland, by HM Industrial Pollution Inspectorate). The total of individual plant limits for each generator is equal to their bubble. There is, therefore, some limited flexibility within each year for generators to negotiate increased quotas for certain stations in return for corresponding reductions in the quotas of other stations owned by the same company.

Sulphur dioxide emission limits are currently under review in discussions in the UN Economic Commission for Europe. There is every prospect that the outcome of these discussions will be a tightening of national targets.

The Director General of Electricity Supply is not of course responsible for overall environment policy, or for whether or not targets should be set for the reduction of sulphur dioxide emissions, or how large the overall reduction in emissions should be. These are properly matters to be decided, or negotiated internationally, by the Government. However, his duties both to promote competition and to protect consumers mean that he has a significant interest in the way that the sulphur dioxide regime in Great Britain operates. He is concerned to ensure that the regime does not place unnecessary obstacles in the way of competition in generation, and that it permits the objective of reducing the sulphur dioxide emissions of electricity generators to be achieved in a least cost way, so that customers do not have to pay unnecessarily high prices.

In this context it is an important advantage that the UK National Plan on sulphur dioxide emissions specifies ends and not means. Within its bubble, and the quotas for individual power stations, a generator can choose between at least four options in devising a compliance strategy. These are:-

- to rearrange the operating pattern of existing stations so as to make maximum use of non-fossil or low sulphur generation. In practice, the possibilities here may be limited, since the main non-fossil stations - nuclear and hydro - tend already to be operated to the maximum extent feasible;
- to substitute low sulphur for high sulphur fuel. In England and Wales, this might increase the attractiveness of burning low sulphur imported coal over high sulphur domestically produced coal;
- to construct and use new generating capacity. Emissions from new capacity are not part of a company's sulphur dioxide 'bubble', though the legislation requires that new stations have to be designed so as to minimise sulphur dioxide emissions employing what is known as 'the best available technology not entailing excessive cost'.
- In England and Wales, 4875MWs of new generating capacity employing combined cycle gas turbines (CCGT) have been constructed since 1990 and a further 18172MWs are under construction or planned. There are several reasons for this 'dash for gas'; one of them is a perception that gas generation has lower environmental risks and costs than coal, both in terms of sulphur dioxide and carbon dioxide,
- to retrofit flue gas desulphurisation (FGD). Under the UK National Plan, National Power and PowerGen between them have a commitment to retrofit 8 gigawatts of generating capacity with FGD. The capital costs of retrofit are high, and the efficiency of the power station is reduced and its costs increased. FGD is not an unmixed blessing in environmental terms since it requires large quantities of limestone, the quarrying of which may be controversial in environmental terms. Not surprisingly generators have wished to consider how far they can use other options to stay within their sulphur dioxide bubble, before making new commitments to retrofit FGD.

There is, therefore, a degree of flexibility within the sulphur dioxide regime to enable objectives to be met at minimum overall cost. There are, however, some areas of possible concern, One of which involves the relationship between the regime in Scotland and that in England and Wales.

The required reductions in sulphur dioxide emissions are measured against a baseline of the level of emissions in 1980. But the baseline in Scotland is at a low level for a number of reasons, including the fact that the main coal burning power station in Scotland, at Longannet, uses coal from an adjacent deep mine which has about the lowest sulphur content of any coal produced in Europe.

Given the arbitrarily low baseline, the Scottish companies have argued that the removal of sulphur from Scottish production is not necessarily the most efficient way for the UK as a whole to go about meeting the overall national requirement. They argue that because of existing low levels of sulphur emissions in Scotland, the cost per tonne of removing sulphur dioxide there is significantly higher than in England and Wales, and that it would be cheaper in overall UK economic terms if they were to pay for the removal of sulphur from English emissions than for the removal of a corresponding tonnage from emissions at Longannet. They argue that it is a nonsense that, late in 1992, output from Longannet had to be constrained in order to stay within emissions limits, and that its output was replaced by output from stations in England and Wales which emit more SO_2 per unit of electricity than Longannet does.

In a market situation, one might expect that it would be possible to test the contention of the Scottish companies that there are cheaper ways of reducing sulphur dioxide emissions than by requiring them to retrofit FGD at Longannet. However, their contention remains untested because the present sulphur dioxide regime does not allow the trading of emissions consents between companies, or across the Scottish border.

There are other circumstances where inflexibility on emissions consents may have adverse consequences for competition and consumers.

Earlier this year, the Director General issued a statement concerning proposed plant closures by National Power and PowerGen, in the light of reports which he had commissioned on the closures by an independent assessor. In this statement, he made clear his desire to see new entrants into UK electricity generation. This would include the purchase by third parties of plant which the major generators wished to close, or which they wished to sell for other reasons, such as a desire to reduce exposure to the UK market. He commented that some of the widespread concerns about generator dominance could be alleviated if voluntary sales of plant were possible. [1]

However, the sale of existing coal fired plant will not be possible unless the new owner can secure a consent for sulphur dioxide emissions at a high enough level to enable him to operate the plant as he would wish. The purchaser may wish to operate the station differently from the way the seller has done in the past.

Indeed this difference may be the underlying rationale for the sale. In these circumstances, it may not be sufficient simply to transfer to the new owner the quota associated with the station under its previous ownership. However, some modification of the quota may be appropriate, perhaps balanced by corresponding modifications in the limits applied to other generators.

Tradeable permits on the other hand might in principle enable the parties to test the value that each placed on permission to emit sulphur dioxide. However, tradeable permits will only assist in resolving the problems of buying and selling existing generating plant if an open competitive market exists in the permits. If all permits were held by existing generators, for example, this might enable them effectively to deny potential new entrants access to the market through the purchase of existing plant.

A third area of potential concern to the Director is the impact which the tightening sulphur dioxide regime may have on prices in the England and Wales electricity pool, and ultimately in contracts. The electricity pool is the day-ahead wholesale market in electricity. Virtually all electricity in England and Wales, including imports from Scotland and France, is bought and sold through the pool. Generators bid in prices for individual 'gensets' for each half-hour of the following day. These are then ranked in price order and matched against a projection of demand. A price is set by the most expensive generator scheduled to run in any given half-hour. To this price is added further elements to reflect the level of demand in relation to available capacity, and constraints in the transmission system.

There is little evidence yet that sulphur dioxide emission limits have impacted directly on the pool price. However, the main impact of reductions in the limits have still to take effect. The outcome of present discussions in the Economic Commission for Europe may be that the limits will be reduced further and more quickly.

As the constraints tighten, it is likely that the bidding strategies of National Power and PowerGen into the pool will be increasingly influenced by the need to make most effective use of each company's 'bubble' and of such flexibility as exists to move quotas from one station to another within the bubble. The impact may be felt in various ways. For example, the emissions constraints may influence which station sets the system marginal price. And, at times, the constraints may reduce total availability, which will push up the capacity element of pool prices.

In the longer term, emissions constraints will require further investment by generating companies, either in FGD retrofit to existing coal plant, or in new CCGT stations. It would be neither surprising nor inappropriate if the costs of complying with sulphur dioxide constraints found expression in higher electricity prices. But it would be of concern if price changes were amplified or distorted by an inability to reallocate emissions limits so as to reduce costs or to facilitate competition, or if decisions were made on the overall level of environmental regulation without an understanding of the implications for electricity customers.

In conclusion, I have set out the responsibilities of the Director General of Electricity Supply with respect to environmental issues. I have illustrated the sorts of issues with which this confronts us by reference to, first, energy efficiency - where many of the issues have to be reconciled internally, for example in our consideration of price controls - and sulphur dioxide, where the emissions regime is the responsibility of others.

We are very conscious of public concern over the environmental impact of the electricity supply industry. We recognise that these concerns are if anything likely to grow, and that tighter rather than looser environmental controls on electricity generation are likely to result. As regards carbon dioxide and global climate change, the industry is likely to face increased pressure both to improve the efficiency with which it converts primary fuels into electricity delivered to the customer, and the effectiveness with which the customer himself uses electricity. This pressure may well be backed with fiscal incentives - VAT is to be imposed on domestic electricity from next April and discussions are continuing on an EC carbon tax. As regards sulphur dioxide, our concern is that the emissions regime should be designed and operated in a way which permits the most economic solution to the UK's obligations to reduce emissions.

Competition itself can play an important part in securing environmental objectives at least overall cost. It is important that the further development of environmental programmes to meet the challenges of acid rain and global climate change do not have the effect of restricting the growth of competition.

Note

1 Since this paper was delivered, the DGES has secured undertakings from National Power and PowerGen that they will endeavour to dispose of 6GW of coal or oil-fired plant to third parties by the end of 1995.

Discussion: Problems of multi-regulation

George Yarrow

Most basic accounts of regulation focus on situations where a single regulator or regulatory agency acts to change market behaviour in order to promote some or other policy objective. In practice, however, there are frequently many different regulatory agencies making decisions that can affect a given market or industry. To take a simple example, an energy regulator might seek to control an electric utility's prices to prevent abuse of market power, while an environmental regulator might be concerned with controlling the emissions of waste gases. In such a situation the utility will face regulation on two fronts and its behaviour will be affected by both regulators. This can give rise to a variety of problems of multi-regulation connected with the assignment and delegation of regulatory tasks within the political system. For example, it is quite possible that there will exist circumstances in which different regulatory agencies pursue conflicting objectives.

In respect of energy and environmental policies, multi-regulation is complicated by the fact that it is generally not the case that one regulatory body deals with issues of competition and price control and another regulatory body deals with environmental issues. That is, we do not typically find a very simply assignment of responsibilities. Sometimes a regulatory agency will be given, or will take upon itself, a range of different responsibilities - US regulatory commissions have been much concerned with environmental issues, for example - and sometimes a particular set of regulatory responsibilities will be divided among two or more agencies (as when the control of monopoly is divided between a specialist regulatory body and the institutions of general competition policy). The resulting policy structures can be byzantine.

The papers by Wolgang Pfaffenberger and Leigh Hancher are concerned in different ways with the implications of multi-regulation in the energy sector (Pfaffenberger presents a wider evaluation of the administrative approach to environmental policy in Germany, but my remarks here will focus on the *structure* of German regulation). Both examine the division of regulatory responsibilities among different levels of government, which, in line with Hancher, I will refer to as *vertical* multi-regulation, and which can be distinguished from *horizontal* multi-regulation where responsibilities are assigned among bodies at a similar level of government.

Pfaffenberger begins by outlining the relevant institutional framework in Germany and lays stress on the federal structure of the German political system. This leads to a particular structure of multi-regulation, with policy responsibilities being divided among the federal, state and local levels of government. The situation in this respect is very different from the much more centralised regulatory structure to be found in the United Kingdom, and it seems clear that the differences in political structures between the two countries have very different implications for industries such as electricity supply. It is difficult, for example, to imagine anything like the structural and regulatory reforms that were implemented in the electricity supply industry in Britain in 1990 being adopted in Germany over any comparable span of time: the political structure would simply not allow it. (Whether this is a good thing or a bad thing is a matter that I will not consider here).

One fascinating example of the importance of political structure is the role of local government in the German electricity industry. As Pfaffenberger explains, local governments have strong interests in the energy business because they can tax energy sold and use the income to support local services. Hence, they have a strong vested interest *against* conservation measures that would hit energy sales and reduce tax yields, and it would be interesting to hear more about their role in the process of regulatory change. There is, of course, already a rich vein of work on the implications for regulation of the US federal structure, but it would be interesting to see much more work on the German regulatory experience, particularly given the increasing importance of forms of federalism in Europe.

Hancher's paper is also concerned with the vertical division of regulatory responsibilities, but her focus is on the allocation of tasks between the Member States of the European Community and the Community institutions themselves. Adding the European dimension to the political structure described by Pfaffenberger gives a four-level vertical structure in Germany - with regulatory powers assigned to each of the local, state, federal, and Community levels - and obviously increases the number of institutions that can regulate economic activity.

The Community dimension does more, however, than add another hierarchical layer to the regulatory structure. Additional complexities arise, for example, from the fact that enforcement of Community regulations is generally entrusted to the Member States, and there is a further set of multi-regulation issues associated with the division of responsibilities among parts of the European Commission itself. Hancher notes that policies relating to energy and environmental matters require some degree of co-ordination between Directorate-Generals XVII and XI, and comments that 'This aim is of course easier to state than to put into practice'. That understated comment can be generalised to most other aspects of multi-regulation: the political structures through which regulation is supplied are not usually fine-tuned for the promotion of economic efficiency in energy markets (or any other market for that matter), and the scope for inflicting damage on economic performance by means of inappropriate (or downright bad) regulation is both substantial and ever present.

To end on a more positive note, Hancher structures her paper around four questions which provide a very useful checklist for the analysis of multi-regulation in a variety of contexts. Given the objectives of the paper, the questions are framed in a normative way, but they can be simply turned into questions of fact, as follows:

- Who regulates?
- What is regulated?
- What instruments of regulation are used?
- How is regulation enforced?

This short checklist can generate a useful first map of the regulatory structure of interest. In particular, the answer to the first of these questions - who regulates? - should serve as a constant reminder that multi-regulation is the norm, not the exception.

INDEX

The manufacturer's authorised representative in the EU is Springer Nature Customer Service Centre GmbH, Europaplatz 3, 69115 Heidelberg, Germany. If you have any concerns regarding our products, please contact ProductSafety@springernature.com

Printed and bound by CPI Group (UK) Ltd, Croydon, CR0 4YY
13/10/2026
02232373-0016